The Spiritual Purpose of the Physical Universe

As Revealed by Science and the Edgar Cayce Readings

Jeffrey L. Imes

Contents

Preface 1

The Teachings Of Selected World Religions 9

Religious Attitudes Toward Science 19

The Challenge Of Religions 26

The Philosophy Of The Edgar Cayce Readings 34

Are Religion And Science Compatible? 42

What Or Who Is God? 53

Souls – Portions Of God 116

The Purpose Of The Universe 145

God Moved - The Creation Of The Universe 155

Creation Or Evolution 238

Soul Incarnation And The Rise Of Man 253

Physical Consciousness And Causality 282

Christ Consciousness In Action 327

The Opportunity Of A Lifetime 341

References 362

Scientific Notation 370

Edgar Cayce And The A.R.E. 371

About the Author 372

I want to know how God created this world. I am not interested in this or that phenomenon in the spectrum of this or that element. I want to know his thoughts. The rest are details.

- Albert Einstein

Hence we find worlds, suns, stars, nebulae, and whole solar systems MOVING from a first cause.

[Edgar Cayce reading 262-52]

If the Bible does not teach science, among other things, what does it teach? The way to salvation; once you realize that the Bible does not purport to be a textbook of science, the old controversy between religion and science vanishes.

- Georges Lemaitre

Preface

Why did God bring the universe into being? Did the inspiration to create spring from whim or purpose? Was he excited as he beheld the maelstrom of energy that swirled, pulsed, coruscated, and showed hints of the grandeur and vastness that it held in potential, awaiting the right conditions, the right impetus that would allow it to unwind like a coiled spring and launch the fabric of spacetime? From God's point of view, was creating the universe like creating one of those fantastically complex arrangements of dominoes that fill a room or many rooms awaiting a first impulse applied at the right place with just the right force? God moved, gave a small push to the first domino, and all of the rest of the dominoes of his universe started falling, quarks crashed into quarks, protons and neutrons formed, electrons become bound to nuclei, atoms raced away from each other riding the inflationary expansion of the spacetime continuum to eventually be roped in and drawn back together by gravity. Was the coalescence of stars out of a nearly uniform initial distribution of matter, their ignition as new points of light in the vast dark universe, and their explosive death as supernova or cyclical death throes as red giants simply part of a gigantic cosmic fireworks display constructed for his personal enjoyment? Is the natural beauty of the earth and the night sky with its millions of points of light God's three-dimensional equivalent of a masterpiece painted by a renowned artist? Perhaps the vastness of space really just exists as a background for Planet Earth; a glorious nighttime tapestry he prepared for the viewing pleasure of his human children as they go about their lives on the surface of an obscure planet in a remote corner of the vast expanse of spacetime. Maybe the universe has a more profound purpose. Perhaps the universe is the essential and functional product of an ongoing 13.8 billion-year-old evolutionary process he set in motion as part of a higher spiritual purpose to prepare a place for soul development through human habitation and that incidentally has a majestic and beautiful aspect to it.

Science and religion are fully compatible and will eventually converge on the truth about the nature of the universe, the purpose of life, and the relationship between man and God. Science seeks to explain the laws of the physical universe by following a logical and repeatable methodology to verify and correct its observations and theories. Physical science strives to understand the forces that drive the evolution of the universe from the original mix of energy and subatomic particles to the atoms and elements that form matter, planets, solar systems and galaxies. Biological science seeks to unravel the secrets hidden in the miracle of life that blossoms and flourishes on the earth and to discover the nature of mind and its connection to the body. Religion is more concerned with the relationship between man and God, the nature of the soul, and the destiny of consciousness after physical death. The spiritual philosophy of the Edgar Cayce readings brings science and religious thought together as a unified whole. The philosophy reveals that the universe was intentionally created out of spirit as a causal environment for the rehabilitation of souls that have lost God consciousness through selfish activity. Causality in the physical universe enforces the spiritual law that a soul must reap what it sows in the physical environment, which allows the soul to better understand itself and discover its relationship with God.

Science studies this physical facility constructed for soul rehabilitation without understanding its origins or purpose. Just beyond its peripheral vision at the extreme limits of its instruments, science is approaching the boundary between the physical and spiritual realms.

Where science and the Edgar Cayce philosophy intersect, they agree to a remarkable degree of accuracy, but the philosophy also contains much that is currently outside the realm of science and not supported by current scientific evidence. The philosophy encompasses and overlaps with various forms of religious thought. Traditional religion would be better served if its adherents concentrated on the study of the relationship between man and God through open-minded reading and study of various scriptures and the daily use of prayer and meditation for spiritual attunement and guidance. Adherents of some religious denominations spend an enormous amount of time and spiritual energy in misguided attempts to discredit science under the guise of protecting the faithful from secularism or to validate and endorse scripture as the literal infallible word of God. This effort to concoct various formulations of "biblical science" to counter imagined dangers from secular science is a misuse and waste of spiritual energy. This book is primarily written for scientists who seek a way to understand how their study and discoveries of universal physical laws fit within the context of a higher spiritual purpose for the universe and for church members disillusioned with sincere but misguided church authorities and believers who insist that the Bible is a science textbook and that their particular interpretation of the Bible is absolute truth.

The philosophy of the Edgar Cayce readings is not a formal philosophical system derived from the application of logic to the questions of the existence and nature of God or the purpose of the universe and the nature of man and his relationship with God. It is an informal collection of ideas and understanding derived from a large body of readings given by Edgar Cayce over a period of more than forty years in response to requests for information from individuals concerning physical ailments and past soul incarnations. A reading is simply the written record of a session during which Edgar Cayce relaxed, set aside his conscious mind and everyday concerns, and allowed his subconscious soul mind to be receptive to questions given him by an individual desiring physical or spiritual information. Cayce described this receptive state as like being in a large library where the custodian or guardian received his request for information, went among the shelves to locate the appropriate book or material, and brought it to the mind of the waiting Cayce. The library was described as a vast storage facility containing a collection of all of the past and ongoing soul activity on the earth. Everything that we think, say, and do in this world is recorded so that the soul may remember its spiritual successes and failures while active on the earth in relation to its current concept of God and its conscious awareness of its relationship with God.

In a nutshell, the readings describe physical matter as the activity of spirit, biological life as the activity of mind-directed matter and higher-order human life as the activity of will-directed biological life by incarnated souls. The readings declare that electromagnetic energy is a physical expression of spirit, the substance that, along with mind and will, is God. Electromagnetic energy, the main ingredient of the physical universe, is the basis of all matter and in its free form is always in motion, always moving at a fantastically high constant rate of speed through the spacetime continuum. As the readings state, electromagnetic energy is spirit in motion. The creation of spacetime and injection of electromagnetic energy as the physical expression of spirit marked the beginning of the physical universe. Under the right conditions of temperature, pressure and density this energetic equivalent of spirit can condense into matter. Matter is concentrated energy. Everything in the universe that has mass and volume or structure is composed of various arrangements of concentrated energy called atoms. After about 9.3 billion years of physical evolution (the sequence of physical processes that formed matter and material objects), millions of planets had formed. One of those planets had orbital and temperature characteristics favorable for retaining water, the appropriate minerals, and a developing atmosphere necessary for carbon-based life forms. In due course, some atoms present in the earth's environment would become arranged into basic organic molecules that could be organized into cells capable of extracting energy from their environment and reproducing, but this alone was not sufficient for self-awareness or coordinated cellular activity. The Mind of God gave the impetus and the rudimentary self-awareness needed to allow these molecules to work in cooperation as the first unified cellular organisms, and biological life began. This was an intentional alteration of the physical evolution of the universe through the activity of mind, for the purpose of initiating biological life so that through the process of biological evolution (survival and natural selection) an environment capable of supporting higher-order beings could develop from physical matter.

These first biological organisms fulfilled their purpose by extracting energy from their environment to sustain life, by reproducing so that life would have continuity, and by evolving in response to changes in their environment so that life would not remain static. These basic biological activities allowed life to become abundant, diverse and ultimately able to create higher order organisms of more complexity. After about 4.5 billion years of flourishing and maturing life forms, evolution had led to the development of a higher-order human life form. Pursuant to the original purpose for the physical universe and biological life, God stepped in a second time and bypassed the natural ongoing biological evolutionary processes. Using the existing human forms as a pattern, he created a slightly modified and refined human life form whose descendants could be used by souls who had lost God consciousness to regain their awareness of God. This marked the second major alteration of the natural evolutionary processes and its effect was to allow the will and mind of each soul to conduct physical activity in the earth's environment through a complex and advanced biological life form. The new human form would offer incarnated souls a way to express themselves in the causal environment of spacetime where thought-directed actions have observable and direct spiritual and physical consequences. The experience of causality was essential if lost souls were to recognize the detrimental effect of their selfishness and be restored to their original intimate relationship with God. There are many fascinating details within the 9.3 billion-year sweep of history that led to the creation of the first biological life, and within the 4.5 billion-year sweep of history that brought biological life to the diversity and complexity needed to support a human life form directed by an incorporeal and intangible soul.

> *Hence man was given that which is NOT a part of mineral, vegetable or animal kingdoms; though man, by man, is considered of the animal kingdom. WILL - with the environmental forces and the spiritual negative in the serpent - acceded to desire, to become and to experience IN that kingdom of influence. [Edgar Cayce reading 281-54]*

Many of the readings describe soul activity in the distant past on continents that include those now submerged deep in the Pacific and Atlantic oceans and describe large human migration patterns across the earth that occurred tens of thousands of years ago in response to cataclysmic earth upheavals. Ancient man was on the move and in larger numbers than we might imagine. Science has verified few of these ancient earth changes and human activities, and has only recently developed techniques to analyze ancient DNA that are starting to shed light on the movement and mixing of ancient populations. The historical aspects of how Edgar Cayce realized that he had the ability to access records of the past spiritual and physical activities of incarnated souls and when he began to give readings to aid individuals is well documented and will not be covered herein (Sugrue, 1997, Millard, 2007, and Bro, 2011). The more than 14,300 readings Cayce gave during his lifetime touch on many topics, including concepts of holistic health and physical exercise, the origin of the soul and purpose for soul incarnation on the earth, the application of meditation and prayer for greater spiritual attunement of the soul to its creator, the purpose of the universe in relation to soul and man, and the spiritual need for reincarnation. The accuracy of the enormous quantity of information that came through Cayce has not been comprehensively and rigorously investigated but, of the many topics he touched upon, the suggestions related to healthcare have probably been studied and applied more than any other topic (Reilly and Brod, 1988). His nontraditional approach to healthcare and healing has earned him the title of "Father of Holistic Medicine," a strong endorsement and testament to the efficacy of the sometimes unorthodox medical advice that flowed from the subconscious mind of Cayce and the adoption and incorporation of many of his medical suggestions into modern holistic medicine.

This book draws from current scientific theories and facts, the philosophy of the Edgar Cayce readings, and certain Christian religious beliefs as related to science. Specific readings that pertain to the nature of God and relate to the current scientific knowledge and understanding about the creation of the universe, the earth, and the life it contains are interwoven with the text. It contains a collection of thoughts and ideas about the nature of God, the nature of the soul, and the reason God created the physical universe and biological life. It describes how those concepts relate to the current scientific knowledge in cosmology and the evolution of the physical universe. It attempts to correlate information from the readings that mention soul appearances on the earth before about 10,000 BC with the scientific understanding of early man and describes the complicated process whereby the soul and body perceive our environment. And lastly it looks at Jesus and what it meant to be the Christ and why the concept of the Christ is so important to individual souls.

The information relating to the creation of souls and the physical universe and the awareness of man as an incarnated soul encased in a physical body are of most interest herein. There are many other books that touch on these and related ideas or explore them in depth, and there are books that discuss many other topics mentioned in the readings that are not presented herein. The reader is encouraged to seek them out and explore them. The guiding principle of the Association for Research and Enlightenment (A.R.E.), the organization that protects and makes the readings available to all who seek to explore them, is that research must come before enlightenment. The organization actively researches many subjects mentioned in the readings, such as spiritual philosophy, the soul and reincarnation, personal spiritual growth, intuition, healing and holistic health, ancient civilizations and their influence on today's world, and dream interpretation. The readings are vast and deep and the structure and grammar can be a bit intimidating. One can spend a lifetime unraveling and deciphering the various themes that thread their way through them, but the journey is well worth the effort. The references provided herein and the information held in trust at the Association for Research and Enlightenment can be used as a spring-board to a lifetime of rewarding and fulfilling study.

The Teachings Of Selected World Religions

Man probably has always intuited or believed that there is something greater and more powerful than himself, some controlling influence in his life that cannot be seen or explained by his physical surroundings. He has searched for this ultimate authority he calls God and sought a better understanding of this God since he first appeared on the earth. Religions have been established to promote belief systems that have arisen from man's theological studies of various prophesies, scriptures, and attempts to formulate moral codes of behavior. There are now over 4,000 religions worldwide, with Christianity, Islam, Hinduism, Buddhism, Sikhism, and Judaism being the largest groups in descending order of size. There are many similarities among them, especially in the realm of moral behavior, but there are also some significant differences.

Christianity evolved out of Judaism when Jesus was accepted by many Jews and Gentiles as the Messiah whom the Jews were expecting to come. It teaches that there is one God who created heaven, earth, and the universe, and who sent his Son (the Messiah) to the earth for the salvation of the human race, which was mired in self-serving sin and had lost awareness of its true origins. Jesus taught that everyone should love God with all their heart, mind, soul, and strength and should love their neighbor as much as they love themselves, even love and forgive their enemies. These two commandments have roots in the Jewish scriptures which form part of the Christian Bible. Salvation comes to

Christians by repenting of their sins, asking forgiveness from God, and turning from their evil ways. Jesus did not write or dictate any scriptures but left his life and oral teachings as a testament and witness to his authority from God. The New Testament of the Christian Bible is mainly a record of his testimony as compiled and written by his closest disciples and later followers. At the end of his ministry, Jesus was led to death on a cross by Roman soldiers under Pontius Pilate at the instigation of the Jewish religious leaders because his teachings were perceived to threaten their authority and were viewed as blasphemous in their self-righteous minds. After death he resurrected, was restored to life in a glorified body, and ascended to heaven to sit at the right hand of God and rule with him. Those who repent of their sins, seek to live a moral life based on his commandments, and faithfully serve their fellow man will reside in heaven after death.

Islam, meaning submission, is a religion centered on belief in one God (Allah) and the Qur'an, scriptures revealed to Mohammed, the founder of Islam. It teaches that Mohammed was the last prophet God sent to humanity and will ever send to humanity. It shares some beliefs with Judaism and Christianity because it traces its history back to the patriarch Abraham, and ultimately to the first prophet, Adam. Islam does not accept that Jesus died on the cross or was the Son of God. Mohammad began preaching by criticizing the people of Mecca for their practice of idol worship and urged them to return to the message of the prophet Abraham, to the worship of one God. After being rejected and ejected by the citizens of Mecca, Mohammed moved north to the town of Medina where he was able to consolidate his power and influence for the next ten years before returning to Mecca to establish his religion there by force. The main message of Islam is that there is no other God but Allah and that people should lead their lives in a way that is pleasing to Allah.

Hinduism teaches that a human being is not just a body and mind, but also includes a spirit or the spark of God within a soul that works through the body. Its teachings are recorded in sacred texts known as the Vedas. Truth is eternal, is the only reality, and is the expression of the one God. All persons should strive to live a life devoted to right conduct, righteousness and morality. The goal of the individual soul is release from the ongoing cycle of death and rebirth by uniting with God through the realization of its true nature. The idol worship so prevalent in Hinduism arises from the belief that God manifests himself in his creations in many ways, including in human form, and that these manifestations can be prayed to for help and protection. Each idol represents a particular manifestation of God. Hinduism teaches cause and effect and that we are each responsible for our actions and the conditions in which we find ourselves.

Buddhism originated from the desire of the Indian prince Siddhartha Gautama to understand the meaning of life and reason for human suffering. He spent many years praying, meditating, and fasting until he believed he understood the basic truths of life. Siddhartha concluded that God does not exist and taught that there is no soul or permanent essence of an individual self that survives death. Instead, we are only a collection of changing characteristics or attributes that define our self or individuality. His core teaching was that life is impermanent and always changing, and therefore, a life based on possessing things or controlling persons won't make you happy. His Middle Way is a guide to proper living that includes moral codes such as mindfulness, compassion, honesty, service toward others, and avoiding harmful actions toward others.

Sikhism is a monotheistic faith with emphasis on a person's relationship with God and the Sikh community. God is without form or gender. He is directly accessible by everyone, and everyone is equal before God. The ideal of a good and moral life is achieved by meditating on God, manifesting belief as good actions within the community, and being honest and caring toward others. Religious rituals and superstitions are deemed empty and are perceived to have no value. Liberation from materialism and repeated cycles of birth, life, and death is accomplished by overcoming lust, covetousness, greed, anger, pride, and attachment to things of this world. The words of wisdom from Guru Nanak Dev Ji, the first teacher (guru) of Sikhism, and the nine successive Sikh teachers who followed him has been compiled into a holy book called the Guru Granth Sahib Ji, which is considered the final authority and last guru of the religion.

Judaism focuses on the relationships between a unique Creator, mankind, and the land of Israel. It teaches that God is singular, unique, exists without a body, and can perceive the thoughts of men. Jews believe in the coming of a future Messiah and in the resurrection of the dead. Abraham is the founding father of the Covenant, the special relationship between the Jewish people and God, a belief that gives the Jews a unique position as people chosen by God to set an example of holiness and ethical behavior for the world. Judaism does not have a dogma or formal set of beliefs that one must hold. Actions are far more important than beliefs. According to Orthodox Judaism, these actions include 613 commandments given by God in the Torah as well as laws instituted by rabbis and long-standing customs. Judaism incorporates a range of different beliefs, ideas, and values that are known as God's guide to a proper life or more commonly, "the way of life." Jewish beliefs are summarized in Rabbi Moshe ben Maimon's thirteen principles of faith.

These six formal religions have been around for several hundred years to several thousand years (Sikhism 500 years; Islam 1,400 years; Christianity 2,000 years; Buddhism 2,500 years; Hinduism 4,000 years; and Judaism 4,000 years). Does the fact that so many humans have spent such a significant part of their time on the earth seeking answers to questions about the existence of God or the nature of the soul indicate that there is an inherent and deep-seated motive and purpose within the species we call Homo sapiens to discover its relationship with God? Is there an ever-present but largely hidden or suppressed inner drive to seek a higher meaning for life that keeps forcing itself into our consciousness and demanding our attention? Many Christians who are tolerant of their Christian neighbors that adhere to the teachings of different denominations have no qualms expressing the belief that only Christians, defined in a narrow-minded manner as those who believe in Jesus, will make it to heaven. Why do many Christians find it so easy to believe that God does not care about or have a plan for a child raised in India or China who has never had the privilege or honor to hear about Jesus or to read the Bible? Is there any validity to this belief? Is it true, false, or perhaps partly correct? This attitude is not exclusive to Christians; many adherents of other religious beliefs also feel that the religion of their youth and upbringing is the preferred or exclusive path to enlightenment or salvation or heaven. Christians who would feel uneasy judging their neighbors for their shortcomings and failure to live up to the high standards that Jesus set and would willingly open their minds and hearts in forgiveness for them often have little sympathy for those not called Christian.

> *Now do not mix religious thought with religion, nor Christianity, nor brotherly love, nor that of Confucianism, or Brahmanism, or zoism [Taoism], or any ism, but rather that of the awakening of the individual to the truth of the divine heritage in each individual that may respond to that of creative energy in the universal forces of EVERY thought, and that may be applied in the workaday, the material, the spiritual, the social, the EVERYDAY life of each individual, whether as to its association with its OWN household or its neighbor. [Edgar Cayce reading 254-48]*

In response to the question as to whether the faith of a Buddhist or Muslim is equal in spiritual value to faith in Jesus the Christ, the source of the readings stated that the sincere seeker who receives a prophet "in his name" will receive a reward in accord with the thought and feeling that is aroused in the mind by that name and the spiritual activity that is manifested in the seeker's life as a result. Each religion may have a role to play in the development of human spirituality in the time and place of its influence, but all religions are only stepping-stones on the path through those experiences that will eventually awaken the seeker's consciousness to the understanding that the Son of God demonstrated the way and that the Christ is the advocate with the Father (262-14). This statement doesn't mean that Christianity is the only way to God, but it does mean that the two commandments that Jesus gave to the world, not to Christians, are ultimately the only way whereby a soul can come to the knowledge of God and return to communion with him. The commandments of Jesus are the rules that define the ideal behavior and material expression of incarnated souls. They are the instructions for personal and societal behavior for those who would seek God. They are firmly rooted in love. The value of the commandments to the soul who applies them transcends all ritual, dogma, moral codes, and teachings of any and every religion or philosophy devised by man. They encapsulate universal spiritual ideals that will lead souls back to God consciousness when applied in the everyday activities of each individual.

How do these various attempts by people all over the world to understand, characterize, define and explain God and our relationship with him compare with the Cayce readings? This book will not attempt to address that far-reaching question by a direct comparison of religious doctrines. The material in this book may help elucidate some the similarities and differences between the readings and the teachings of these various religions to those who are familiar with each religion, but that is not the purpose of the book and was never the purpose of the readings. As reading 254-80 states, "If ye look for differences and still set the differences aside, you'll continue to have differences!" Comparative religion may have its place in the academic world and as an exercise in trying to better understand other cultures and religious beliefs and practices, but it poses a danger when religious leaders or laymen use it as a reason to separate themselves from, or discriminate against, persons raised in or practicing a different religion. Again and again, the readings warn about the danger of creating a new religion, or as is often stated, a new schism, ism, cult, or creed around the spiritual, philosophical, and metaphysical truths coming to light through the spiritually receptive mind of Edgar Cayce (254-80; 254-87; 254-91; 254-92; 254-103;

254-105; 254-111; 263-98; 262-101; 412-13; 954-6; 1549-1; 2787-1).

We find the same contentions arising in that called in the present denominationalism, and each one crying, "Lo, here is Christ - Lo, this is the manner of approach - Lo, unless ye do this or that ye have no part in Him." [Edgar Cayce reading 364-9]

An attempt to create yet another religion based on the Cayce readings would immediately be divisive and would once again lead men to focus on the differences in their belief systems instead of the commonality of their core values. It would also lead to the accumulation of power by and ultimate corruption of the well-meaning individuals who create and promote the new religion just as it has in far too many Christian denominations. The material from the readings was never intended to be pushed upon an unprepared and unwilling world but was intended to be compiled, preserved, and made readily available to any and all persons who come seeking a greater understanding of their relationship with God. The readings make it clear that it is the duty of Christians and all persons of every faith to foster a sense of harmony among men by applying the commandments of Jesus in a practical way in our everyday lives. The teachings of Jesus are presented as a standard for Christians and non-Christians alike. They confirm that Jesus and God have a special relationship but also state that all other souls have the potential to have a similar close relationship with God and every soul should have a shared desire to help humanity to realize its spiritual origins. Jesus and Christ are not synonymous terms or identical entities within the context of the readings. The readings declare that the most important spiritual connection between man and God derives from the activity of the Christ, more accurately the Christ Spirit that resides within the mind of every man and can be activated by conscious effort. The emphasis on the man Jesus in Christianity as proclaimed and practiced by most Christians is not necessarily the same as an emphasis on the Christ Spirit as explained in the readings.

For the religion of Christ is a practical, everyday living, and

not a tenet or a thing to be taken lightly but is to bring life,
purpose, peace and harmony by its application in its relation-
ships to the fellow man. [Edgar Cayce reading 781-5]

Christianity is a community of like-minded people who (should) follow the teachings and life principles of Jesus. Christian denominations generally have a set of common core beliefs that include Jesus as the Son of God and belief that during the first century AD, he taught the Jewish people of Judah and Israel a way of life that emphasized loving God and neighbor above any selfish interests. It is a common assumption or belief among Christians that if we are fortunate enough to have natural or acquired moral strength or faith, if we are fervent enough in our belief in Jesus as a historical figure and Savior, and if we have an acceptable history of high-quality deeds, we will be accepted into a place of eternal abode called heaven, where all will be joyful and we will spend our eternal existence full of praise for God and without want or need. Exactly what this means or implies doesn't seem to be well understood, as is attested by the fact that there are different ideas about the true nature of heaven and many somewhat different ideas about the proper way to get there.

When we die, our bodies will be gone, left behind to decay as all living things do, but what moves on? And do our bodies really remain dead, and if not, how are the constituent atoms reassembled into a functioning resurrected body? Jesus seems to have figured out this perplexing problem, but it may not be pertinent to our situation. Will we really have a need to reconstitute our bodies for earthly life? Why would we when we are supposedly finished with the physical world and on our way to one of the two other possibilities, heaven or hell? Will our body get raised up as some gloriously transformed super body that can float through the heavenly realm? Do we get our hair back or do we remain bald? Is it just our conscious mind that survives death, and if so, why do we think we will act any different in heaven than we did on the earth? Will we be more compassionate or caring after we die than we were before we died? Will we never let an angry word or thought slip out just because we are dead and can no longer communicate through a set of physical vocal cords? There presumably will be no possessions to be arguing over or beautiful persons to desire sexually, but will there never be a moment of jealousy between two individuals just because their minds no longer control physical bodies?

Christianity, depending on the denomination and in similarity to other religions, has a set of implied or written disciplines that are to be used to prepare each individual to meet with or be in the presence of God. Commonly suggested methods for building a closer relationship with God include the reading of scripture (often described by Christians as the literal authoritative infallible word of God), prayer (a form of verbal or non-verbal petition to God), and worship (the action of honoring and glorifying God and exalting him above all else). It might be argued that the three most important disciplines should be scripture, prayer, and meditation (the emptying from self of all that might hinder communion with God), with the last being perhaps the most important but the least used or taught in Christianity.

Scripture, while important, can hamper our search for God when it is interpreted in a dogmatic and literal manner; when it is used to declare others heretical because they dare to have a different interpretation of certain passages or entertain ideas not deemed compatible with certain verses; when it is used to justify actions not in accordance with the two principal commandments of Jesus; when it is used to form an exclusive club that condemns or excludes others; and even when it is used as a science textbook. Prayer is verbally or mentally talking to God and asking him for guidance, help, support, strength, blessings, etc., or giving him thanks, praise, devotion, etc. It is a form of communication to God as opposed to communication from God. It can be inherently selfish in nature if its focus primarily or constantly is to express the perceived physical desires and needs of the supplicant rather than being used as an opportunity to share one's thoughts, fears, worries, and expectations with God and as a means to help bring the individual will into alignment with God's will. Prayer that is constructed around the thought that "I need …", or even "My neighbor needs …" sometimes can place the individual in the position of telling God the best approach to handle a situation or the best way to aid a neighbor.

It is natural to ask God to intervene in our lives in a certain way. We often speak to God in prayer out of an earnest desire to receive healing and relief from pain during a severe illness or as part of an attempt to navigate a serious personal crisis that threatens to overwhelm us. We want to direct him toward a certain outcome that we perceive as more favorable and will lessen the pain and fear that we are feeling. We must remain mindful that God may have a different idea of the best outcome of certain events in our lives and that our ideas may not conform his plan for us. This is especially true because we usually consider the needs of the body, not the needs of the soul. It is easier to pray than to meditate. Meditation requires discipline, dedication, and directed effort. Meditation is the emptying of a person's mind of all that might hinder communication with God for the purpose of allowing God to enter into one's consciousness, or more properly, to become consciously aware of the ever-present God that resides within us. It is a process of becoming receptive to the "voice" of God and patiently waiting and listening for God to speak to us. Its true and highest purpose is to reorient the conscious mind and soul mind into a more spiritual direction that makes us more aware of God's presence and his will and purpose for us.

Religious Attitudes Toward Science

The attitudes of people across the globe toward science vary greatly depending on their level of education and —perhaps surprisingly or not, depending upon your point of view— their identification with religion. The 2018 Wellcome Global Monitor survey, conducted by the Gallup analytics service, questioned 140,000 persons in 140 countries to assess the attitudes of people toward health and science in general with a focus on vaccine acceptance (Wellcome, 2018). Given a very basic definition of science and scientists, and then asked if they understood the meaning of that definition, 83 percent to 94 percent of people questioned in North America, Europe and Australia answered that they understood all or some of it, while only 31 percent to 64 percent of people questioned in Asia and Africa answered that they understood all or some of it. An understanding of science generally is acquired through education and the survey undoubtedly reflects the vast difference in educational systems and opportunities, and in the number of people in different countries who complete at least a basic education. Similar large divergences by region were identified when the question more specifically asked how much the individual knew about science. When this data was broken down into three age categories (fifteen to twenty-nine, thirty to forty-nine and over fifty), the middle-aged and older people in Asia and Africa were much less familiar with science than the younger people, indicating that science education in these areas generally has lagged behind the same in other regions.

One survey question indicated that 76 percent of North Americans identified with a specific religion and 57 percent of those people stated that science disagrees with the teachings of their religion. Perhaps this finding is not too surprising considering the large number of Evangelicals and conservative Protestant Christians in the United States. Only the Southern Europeans and to a lesser degree Western Europeans and Australians were as pessimistic about the compatibility of science and religion, perhaps also related to their generally Christian populations. In these latter three regions, 50 percent or more of religious people stated that science disagrees with their religion. Religious people in Central and East Asia, Europe, and Australia are more likely to believe science over religion when there is a discrepancy, but religious people in the high-income and high-education region of North America, and more specifically the United States, are almost twice as likely to believe their religion instead of science when there is a conflict between the two. Religious people in Africa, South America, and Southeast Asia are also far more likely to believe their religion instead of science when they conflict.

The Pew Research Center is a nonpartisan American think tank that conducts worldwide surveys to gather public opinion statistics on various social issues and demographic trends. Twenty-four individuals in each of three religious groups —Muslims, Hindus, and Buddhists— were interviewed in Malaysia and Singapore by the research center as part of the 2018 Wellcome Global Monitor survey to assess public perception of science in relation to religious beliefs. There were certain variations among the responses to the questions but there were also some common themes that could be identified within each religious group.

The Muslim interviewees generally saw some conflict or tension between science and religion, especially when it came to questions about the theory of evolution and their religious beliefs about the origins of human life. Many Muslims believed that God created humans in their present form and that all humans are descended from Adam and Eve, and some claimed that evolution is only a theory that is yet to be proven true because there is no solid evidence to verify it. The theory of evolution was often described as being incompatible with the Islamic teaching that humans were created by Allah. The focus upon evolution as a point of contention between science and religion parallels the thoughts of many Christians on the same subject. But at least one Muslim said that science does not conflict with Islam; they both are just trying to explain the same things from different perspectives.

The interviewed Hindus generally indicated that science and religion were compatible and that elements of science were incorporated within their religion. They were generally comfortable with the theory of evolution and believed that it is contained within parts of their religious teachings. Many of them had no difficulty expressing the concept that humans evolved from primates. The Buddhist interviewees were in general agreement with the Hindus in perceiving few conflicts between science and religion. They considered science and religion as two separate and unrelated arenas of thought, seeing science as a means of observing and studying physical phenomena and religion as providing guidance for living a moral life. The Buddhists generally saw little conflict between evolution and their religion and believed in the theory of evolution. Some did not believe their religion even addressed the origins of life on the earth and thus believed that science and religion cannot be in conflict because they do not intersect and reside in different or parallel realms.

A general understanding of the attitude of the adherents to the Sikh religion toward science can be extracted from various online sources. The website of the Institute of Sikh Studies headquartered in Chandigarh, India, suggests that there is little to no conflict between science and the Sikh religion and even indicates that the Sikh scriptures recognize a close interdependent relationship between science and religion. Both seek to understand reality using different and complementary methods and both forms of knowledge are valuable. Scientific investigation is largely considered to complement religious thought and a spirit of humility and cooperation among scientists, theologians, and religious authorities is considered desirable and essential. Science and religion are recognized as the two strongest forces that influence human history, and because of this influence, there is an obligation to distinguish genuine science from pseudoscience, identify selfish intentions that harm humanity, and separate informed spirituality and religious thought from empty rituals and superstitions.

A survey by the Pew Research Center in 2014 questioned 35,000 persons throughout the United States about their religious affiliation and their thoughts about various issues. It did not focus on science but did ask the respondents about their views on evolution. The study found that 53 percent of Muslims, 86 percent of Buddhists and 79 percent of Hindus of the surveyed respondents in the United States believed that humans and other living things have evolved over time, with the majority of respondents believing that evolution is a natural process (Pew Research Center, 2014). The responses of those who identified with other religious groups varied, but the responses of Evangelical Protestants, historically Black Protestants, and Mormons were the most strident in their rejection of evolution. The survey found only 38 percent of the Evangelical Protestants, 50 percent of the historically Black Protestants, and 42 percent of the Mormons among the surveyed respondents believed that humans and other living things have evolved over time, with only a very small minority believing that evolution is a natural process.

Christianity and Islam are derivative of Judaism. The first five books of the Jewish scriptures, called the Pentateuch (or Torah in the narrowest meaning of the word), describe the creation of the universe and life on the earth, the exodus of the Israelites out of bondage in Egypt, and the covenant and laws that join God and the Israelites. The Old Testament of the Christian Bible includes these Jewish scriptures as well as books of history and prophesy, whereas the New Testament of the Christian Bible is considered to be a record of the fulfillment of the prophesies of the Old Testament by the coming of Jesus as the Messiah.

The Qur'an of Islam draws heavily from the Christian Old Testament and therefore the Jewish scriptures. The 2014 Pew Research Center survey found that 79 percent of the Jewish respondents believed that humans and other living things have evolved over time, with a majority of the responders (58 percent) believing that evolution is a natural process. But Catholics, mainline Protestants, Evangelical Protestants, and Muslims —all Abrahamic religions whose scriptures rely heavily on the scriptures of Judaism— have entirely different beliefs regarding evolution and the creation of the natural world. They have effectively reinterpreted the Jewish scriptures upon which their religions are founded to create a new worldview. The survey showed that 66 percent of the Catholic respondents believed in some form of evolution, but only 31 percent believed it is a natural process; 64 percent of the mainline Protestant responders believed in some form of evolution, but only 28 percent believed it is a natural process; 53 percent of the Muslim respondents believed in some form of evolution, but only 25 percent believed it is a natural process; and in the most extreme rejection of evolution, 38 percent of the Evangelical Protestant respondents believed in some form of evolution, but only a definite minority of 11 percent believed it is a natural process.

A survey conducted by the Barna Group in late 2011 to better understand why young adults with a Christian background were choosing to leave the Church identified six main reasons (Barna Survey, 2011). Three of those reasons are of interest herein, with one being of primary importance. One of the reasons given by the surveyed young adults was the antagonism that some churches direct toward science and the perception that church authorities feel that they have all of the answers to life and the origin of the universe and are unwilling to consider any other perspective. Fully 35 percent of the respondents felt that Christians too often think that they know all the answers, and 29 percent felt that churches are out of step with the modern world and the advances in scientific knowledge. Another 25 percent thought that Christianity was not only dismissive of science but actually took an anti-science stance. Nearly a quarter, 23 percent, declared they were fed up with the creation versus evolution debate, but many science-oriented young Christians were trying to reconcile their faith with their interest in science and desire to build a career in academia and science-supported industries. Two other reasons for the declining interest of young adults in the church have some bearing or relation to the science issue. The young surveyed people felt that the church is over protective and used fear-based tactics to get their message across, with 23 percent of the respondents claiming that Christians demonize everything outside of the church. They also were resentful of what they felt is the exclusive nature of Christianity and a lack of open-mindedness, tolerance, and acceptance of others among church members and authorities. Young people today desire to find commonality with each other even if that means ignoring real differences in personal outlook or opinion and 29 percent agreed that churches are afraid of the beliefs of other faiths.

The 2020 Barna survey on the changing state of the Christian Church drew from 96,171 surveys conducted over a period of more than twenty years to look at trends in church attendance and frequency of scripture study and prayer in the United States (Barna Survey, 2020). The survey found that the share of practicing Christians dropped in half between 2000 and 2020. The vast majority of that decline occurred between 2009 and 2014. About one half of these people who no longer attend church regularly described themselves as non-practicing Christians, but the other one-half described themselves as non-Christians, which includes atheists and agnostics. Church attendance between 2000 and 2020 declined from an average of about 43 percent to about 30 percent, dropping precipitously between 2009 and 2017. The survey found that the main factors that contributed to the declining interest in traditional churches include the growing number of young people in the United States population, disputes about who qualifies for church membership or who can become a church leader, past and current church scandals, and disagreements over the role of the church in politics. The spiritual lives of Americans are undergoing a major transition as young people become dissatisfied with established religion in general or are seeking something more spiritually satisfying.

The Challenge Of Religions

It seems there are numerous ways to misuse scripture and the trappings of religion, and men and women have discovered all of them to the detriment of their souls and the souls of their fellow human beings. It is unimaginable to many Christians today how creative church authorities were a couple of hundred years ago in using scripture to justify white supremacy and the immoral institution of slavery (Morrison, 1981). Covert racism is still visible just beneath the surface of American society and expresses itself as continued resistance to desegregation in many subtle ways through redlining practices and inequality in the workplace directed against African-Americans and the continued reverence of Confederate statues and flags despite the pain they inflict through their symbolism. Independent and nonpartisan research and statistical surveys conducted by the Public Religion Research Institute during the past decade have definitively shown that white Christian Americans consistently have much stronger and blatant pro-racist attitudes than white religiously unaffiliated Americans (Jones, 2020). It is a sad commentary on American religious institutions that un-churched people are much more racially tolerant than active church members that vocally profess allegiance to Jesus and claim to follow his teachings.

When churches engage in the promotion of self-serving doctrine and immoral activity, the effect of those actions continues to roll through the nation long after those practices come to an end on an institutional level. There is a latent attitude that remains embedded in the psyches of reincarnating souls who, for example, previously enjoyed special privileges within the institution of slavery or who were victims of its cruelty. Those souls bring past attitudes and fears with them into subsequent lives, causing wave after wave of intolerance and racism to persist in American society. History that begins with the evil intentions and selfish actions of incarnated souls, even when blessed by religious institutions and packaged in inspiring sermons that use words like God, Jesus, and love, doesn't end with the death of the original actors. The effect of this unholy activity will unfold through generation after generation until those depraved ideas are purged from the minds of the souls who perpetrated the original evil and the souls that blindly followed them. If attitudes and actions such as these are not purged from the minds of its citizens, then America risks losing its precarious position as a moral force and leader in the application of democracy, tolerance, and righteousness and will enter a decline from which it may never recover. America is not just a land of freedom and opportunity; it is a land of responsibility. Living in the land of religious freedom does not give any individual the right to dictate the religious beliefs and personal morality of others. Freedom does not mean the freedom to force a particular form of Christianity upon others or even to force Christianity upon others through religious bigotry or governmental legislation. America is not to be just a Christian nation but a nation devoted to freedom of worship.

> *These [justice, mercy, peace and truth] are the purposes upon which this land [America] were founded; and they are those forces, those self-evident facts of man's existence that are a part and parcel of every soul's expression in the material plane; that freedom of speech, freedom of the purpose for worship-fulness of Creative forces according to the dictates of one's own conscience, shall never perish from the earth. [Edgar Cayce reading 2167-1]*

The readings specifically warn that if the attitudes and actions of Americans are not kept more consistent with and closer to the principles of behavior commanded by Jesus, then the status of America as a beacon of hope through democratic institutions that respect all persons and through the application of high moral principles in personal relationships will slip away from this country and be offered to another. In the words of the readings, Christianity will "wend its way westward – and again must Mongolia, must a hated people, be raised" (3976-15). These words seem especially relevant today when the Muslim Uighur people of Mongolia are the victims of a brutal cultural genocide perpetrated by the Chinese government and Christians throughout China have been driven underground by sanctioned repression of Christianity. Will a sufficient number of Americans turn back to God and the commandments of Jesus as the basis of all personal and societal interactions in time to save America? This is the major challenge that the American people face today, not the quest for economic security or military dominance in the world or political power, and Christian religious denominations will have a leading role in determining how the future of America unfolds. The recent rise of political leaders who aspire to circumvent the democratic process in pursuit of unaccountable authority, who pander to white supremacist groups and use falsehoods and deceptive messages to deceive voters and sow distrust of the democratic process, and the willingness of many Christians to trade honesty and integrity for political power is not encouraging for the future of America. Perhaps the most disturbing misuse of the Christian religion in America today is the rise of Christian Nationalism, a gross distortion of the teachings of Jesus and the beliefs of the Founding Fathers used by evangelical Christians to justify a blatant grab for political power. Focusing spiritual energy on trying to discredit secular science by formulating and promoting various versions of biblical "science" also does not contribute to the spiritual advancement of America.

> *Is America fulfilling her destiny? (A) ... Is America as a whole? This is as has been given. If there is not the acceptance in America of the closer brotherhood of man, the love of the*

neighbor as self, civilization must wend its way westward ... What have ye done with the knowledge that ye have respecting the relationships of thy Creator to thy fellow man? [Edgar Cayce reading 3976-15]

According to the readings, the sunken continent of Atlantis (Plato, 2008) was the location of a civilization that endured serious internal struggles between forces within the society that desired to remain aligned with higher spiritual principles and forces that were delighted to indulge in every selfish and self-gratifying activity. The spiritual and moral battle between these competing world views as represented in the godly activities of the Sons of God and ungodly activities of the sons of Belial literally tore apart the nation of Atlantis. The abuse and misuse of portions of the population thought to be inferior and the misuse of destructive technology by the sons of Belial led to the eventual collapse of the Atlantean society and to geologic upheavals that eventually destroyed the continent. Reading 5748-6 in 1932 and reading 3029-1 in 1943 state that there has been and will continue to be a greater influx of souls from the Atlantean and other ancient civilizations entering the earth that will be the beginning of the change in the races (probably not meant to indicate skin color). To what extent are the present racial problems and sense of entitlement inherent in American white-supremacy attitudes and movements today a continuation of the same conflicts between the Sons of God and the sons of Belial that led to the destruction of the continent of Atlantis?

The phrase sons of Belial is also used in the Bible (for example, Judges 19:22, 1 Samuel 2:12, and 1 Kings 21:10) and refers to a person who is worthless, good for nothing, wicked, or godless and likely to lead himself and others to ruin and destruction. The sons of Belial in Atlantis had little concern for the rights of others; they abused those they deemed inferior and engaged in activities that fostered an atmosphere of disrespect, distrust, and denigration of others. They attempted to gain control and power over the minds and bodies of those they demeaned by subjecting them to forced labor, servitude, and slavery. This sense of superiority and desire to mistreat and manipulate others may be the greatest threat to American society today and can only be warded off to the extent that the majority of the American people decide to hold, to remain true to the spiritual principles contained in the commandments of Jesus. Will America follow Atlantis to destruction, perhaps not in the form of the physical destruction of the North American continent but a spiritual and moral destruction leading to decadence, economic stagnation or decline, and a divided and divisive political climate with a rise in authoritarianism?

Despite all the good they have done in raising spiritual and moral awareness and as social organizations, religions can be terribly flawed institutions. History contains too many examples where the reason and purpose for the formation of a new religious organization is born out of selfish interests, narrow-minded interpretation of scripture, or suspicious revelation rather than altruism and a desire to bring the knowledge of God to humanity. The real problem with religious organizations is that they are created by men (and occasionally women), and the men and women who they attract are too often driven by emotion and become blind, uncritical followers. Man's penchant for power and the inner smug knowledge that he knows better than his neighbor and has special insight into the truth too often leads him to build holy structures out of corrupt components. When men with selfish motives, intentional or unintentional, decide that they know the truth better than their neighbor; have a higher calling that justifies trampling on the rights of lesser men; and are able to gather a core group of believers around that perverted truth; they can detrimentally affect the lives of many people, even to the point of altering history. When selected self-serving truths and falsehoods are used to push a particular agenda or are used to elevate an egotistical leader to a position of power, the religious structure that arises is a castle built on sand. Unfortunately, the history of religion is rife with men who became corrupted by ecclesiastical power and wealth and position, men with strong self-assured egos who were certain that God gave them special authority and knowledge that they felt compelled to force upon their fellow man.

> *Man is hedged about by beliefs, by cults, by schisms, by isms, - yes. And those things have been created by man that he hath given power in themselves to rule his days. Yet this is only because man has given them such power. For the spirit of truth and wisdom is mighty, and a bulwark of faith and hope to those that trust in Him. But they that give others, other things, power over themselves become subject unto them. Thus hath He declared, as was given of Him [Jesus] who is the way, the truth and the light, the first, yea the whole of the commandment of the Lord is encompassed in this: "Thou shalt have no other god before me, neither in heaven nor in earth, nor in things seen or unseen, but thou shalt love the Lord thy God with all thine heart, thine soul and thine body, and thy neighbor as thyself." [Edgar Cayce reading 2454-4]*

Whenever religious leaders and their followers embrace evil, even in a misguided attempt to do good for humankind, the harmony of the universe is disrupted for generations to come (Griswold, 2021). The ongoing cycles of religious and societal turmoil will only cease when people become seekers of God, not seekers of religion, and when they desire to follow the commandments of Jesus to love God and love neighbor as much as they love themselves. Society will remain in a state of disharmony and division as long as its members choose to follow any doctrines, dogmas, traditions, and or rituals that demean or lessen the value of other members of the society. Too often man seems to desire religion and its dogma and doctrine more than he desires a loving relationship with God. Maybe we just don't understand what God really wants us to do. Jesus seemed to be pretty clear about it, but man keeps adding to his teachings and cluttering them beyond recognition. Jesus didn't set up a church organization, and he didn't attempt to eradicate any existing religion, and he didn't try to reform any existing religion or denomination and use it as a platform from which to promote his message. He didn't try to convert the Pharisee or Sadducee leaders to his thinking so that he might gain access to their organization of believers and the religious power they wielded. He didn't want to organize a new religious institution or hold the corrupting political power that would come with it. He only asked that each and every individual seek and love God with all their heart and mind and consent to treat all other people, no matter what religion they followed, with love and respect. The biblical fruits of the spirit are the rules of engagement for interpersonal relationships that are acceptable to God.

Religions see the big picture but the devil is in the details.

There are many fine people who claim allegiance to various religions today. Members of most religious groups try to be socially responsible through faith-based organizations with various degrees of success. The Salvation Army Shelters, Catholic Children's Aid Societies, and Jewish Social Services are only three examples of the dozens of religion-backed social organizations that attempt to help needy people. These and similar institutions, whether Christian or non-Christian, are putting into practice the directive of Jesus to shelter the homeless, feed the hungry, clothe the naked, and visit the sick (Matthew 25:35:40). In the past few years members of many smaller churches have become more socially conscious and more proactive in meeting the needs of their communities. It is a testament to the basic goodness of most people that even religions and religious denominations that have passed through periods of corruption and decadence still can do so much good in the world, but why should it be necessary to weigh the good done by any religion against the bad it may have perpetrated in the past? The challenge of religions today is to stay focused on, and true to, the commandments of Jesus or their equivalent expression. For adherents of several Christian religions, that may mean to stop wasting spiritual energy and expending spiritual resources in trying to discredit science by promoting creationism and its derivatives or denouncing the theory of evolution or the well-documented fact of a 13.8 billion-year-old universe. These activities do nothing to spread the good news of a loving God and his desire that we treat each other with love and respect, but they do drive educated young people away from church doors.

The main focus of this writing is not to document the unsavory origins and histories of some Christian denominations or other religions but is to focus on some aspects of selected topics within the scientific fields of cosmology, anthropology, and biology that are of particular interest and relevance to the Edgar Cayce readings and some religious beliefs. These are challenging areas for adherents of some, but not all, religions. Many Christians have difficulty reconciling their religious beliefs and the doctrines of their denominations with certain scientific theories and observations. Notable among these are theories of the origin of the universe, the rise of man, and the evolution of biological organisms. Many conservative Christians prefer to believe literal interpretations of a few inconclusive and unverified Bible passages regarding these topics even though most of these ideas conflict with the established scientific understanding of these aspects of nature. The challenge to these modern Christians is the same challenge faced by Christians some 400 years ago when certain Christian authorities insisted that the earth was the center of the universe and demanded that every Christian adhere to that doctrine. Science was clearly in the right on that issue because it sought the truth about the physical universe though observation and investigation instead of biased and blind acceptance of untested ideas. How long will it take conservative Christians today to disentangle themselves from the baggage of false science and move forward to the more exciting arenas at the intersection of science and spirituality: the relation between the creation event of the universe and the Creator, the biological process of evolution and the entanglement of souls in the physical realm, and the physical and mental connection between a human body and an incarnated soul? This is the way forward and points to a future where religion and science merge and complement each other, where the energy expended by conservative church members in antagonism toward science and in efforts to remove certain scientific subjects from school curricula is redirected into more positive, fruitful, and rewarding spiritual pursuits.

The Philosophy Of The Edgar Cayce Readings

The philosophy of the readings has much in common with the major religions but also much that makes it stand apart from them. Unlike Buddhist teachings, the readings affirm that God exists and describe him as a spiritual being, not a person in the sense we normally think of persons, but our relationship with God can be just as personal as we desire. Although God is commonly referred to using the personal pronoun "he," it is only a term of convenience. God does not have a gender. Jesus holds center stage in the readings, but as part of a far greater spiritual presence and activity than accredited by Christian doctrines. "Jesus" and "Christ" are not synonymous in the philosophy of the readings; "the Christ" is associated with the mental and spiritual link between Jesus and God and between man and God. The moral code is similar to the highest and best that can be found in the major religions but does not include specific social rules of behavior, such as is found in the Mosaic laws. The individual is tasked with setting himself the highest standard of thought and behavior he can conceive, using the life and teachings of Jesus as a guide if necessary, and mentally and physically applying himself daily to ensure that every facet of his life meets those standards. The readings agree with Buddhism in that attachment to material goods or power will not lead to sustained happiness but go farther to say that such an attitude blocks the mind from becoming aware of God and inhibits soul growth.

The sweep of history described in the readings goes far beyond the few thousand years of Judaism and Hinduism and outlines a long struggle during which souls who previously lost their awareness and understanding of God while engaging in selfish activity in the spiritual realm have been seeking to rediscover that lost relationship. The process involves learning activities in the spiritual realm designed to refocus and realign the soul mind and incarnations on the earth between these learning activities to demonstrate and explore their current understanding of God as Creator. Like Hinduism, the readings treat soul reincarnation as a fact and as a necessary activity for spiritual growth, but unlike some Hindu beliefs, incarnated souls are said to be currently confined to human bodies and do not have the ability to incarnate in animal bodies. However, the Hindu belief in reincarnation in animal form may be derived from previous soul experimentation with the carnal natural world before the current human form was created. The readings suggest yoga and yoga-like exercises as a beneficial practice to ensure the proper balance between body and mind. They are in general agreement with the Hindu concept of spiritual energy centers located along the spine and in the brain (chakras) and indicate that the endocrine glands are centers of communication between the soul and the human body. Like Judaism and Islam, the readings emphasize that there is only one God, but the readings go farther by repeatedly pounding home the message that God is One and that all that exists is of God and is God.

During the first few years that the readings were given, they primarily concentrated on addressing the physical cause of illness for individuals who requested advice about health issues and were in need of medical attention, often after exhausting traditional medical options (physical readings). Many of the treatments outlined were nontraditional applications of medicinal compounds and light or vibration therapy. Later, the focus of the readings shifted to include the effects of past lives and soul memories on the spiritual and emotional problems being experienced by the individuals requesting readings (life readings). These life readings were often given to individuals who requested information about the purpose of their present life and the cause of the current interpersonal conflicts or financial and business difficulties they were facing. Other specialized readings covered such topics as the development of a series of lessons on spiritual growth, the study of prayer and meditation, and a series on world events during the Great Depression and the eve of World War II. As the source of the readings evolved from giving medical information to describing previous life experiences, Edgar Cayce struggled to become comfortable with ideas that did not always follow the tenants he was taught in the conservative Christian church he had attended during his childhood and adult life. There were times, especially in the early years when Edgar Cayce was learning the extent and limit of his psychic ability, when self-serving persons tried to take advantage of his gifts (Bro, 2011). These questions and problems were eventually resolved in his mind, and he instituted strict procedures to protect himself from self-serving individuals. He was successful in navigating through this minefield without compromising his integrity and remained focused on using his talents to help others. He continued the readings for more than four decades with the understanding that they would continue only so long as they did good and never caused harm to any person.

Edgar Cayce did not promote a cause that he had already consciously determined was truth. He allowed his conscious mind to be set aside so that his personal ideas and beliefs would not interfere with the information he was seeking for the benefit of others. He was careful to seek guidance from God, not from disincarnate souls or other lesser beings that might have their own agenda. This was only possible because Cayce always attempted to live his life to the highest standard he could achieve and unselfishly sought information that would help others, not line his pockets or give him power. He willingly gave of his time and energy for years to help others, did not get rich from his talent, did not promote himself to generate fame or fortune, did not start a church or cult, but he did die prematurely because he would not slow down the pace of his readings schedule long enough to let his body recover from the stress, strain, and wear and tear the practice of conducting the readings imparted to his body. The pressure to answer each of the overwhelming number of requests for medical and spiritual help that flooded into his office each day as his talents became more widely known were a burden on his mind and a strain on his body.

Religions usually promulgate a set of spiritual and moral truths and expect or demand that their members believe them and think and live in a manner that is stipulated as the proper expression of those truths. The readings offer a set of spiritual truths and indicate that any person who is seeking to better understand their relationship with God and who can benefit from the readings will find them, not as a matter of luck, but as a matter of spiritual law. There is no compulsion or stricture that a seeker must believe the history of souls or humanity as outlined in the readings, a human history that extends hundreds of thousands of years before the earliest written historical records and a soul history that extends back to the moment of their creation before time had any meaning. These seekers are directed to study, research, and explore the ideas presented in the readings by putting them into daily practice and come to believe them only as they are proven to be a source of spiritual growth for the individual soul and a force for good in the world. In the words of the readings, we must put research before enlightenment otherwise we are simply the blind leading the blind. We do not learn truth simply by listening and nodding our heads in agreement; there must be an element of investigation and application of the offered principles and a resonance with the spirit within us that bears witness to the truth of those principles.

> *For as ye sow, so shall ye reap - this is an infallible, irrefutable condition or experience. And the more one becomes aware of one's obligations to self, to the fellow man, to the relationships, greater becomes one's abilities to deal with others and to succeed. Hence as has been given, "My spirit beareth witness with thy spirit." This then is NOT a judgment but rather as an ability for an individual who would know himself to check upon himself. [Edgar Cayce reading 257-170]*

As the value of the readings became more apparent and Cayce and others around him became more comfortable with the concept of reincarnation, a central thesis of the readings, a series of readings were taken to better understand the origin and nature of the soul, the purpose for soul incarnation in a material world, and the means by which the soul can restore its lost awareness of its creator and return to a proper loving relationship with God. These readings were used to form the Search for God books that present the core teachings and philosophy of the readings (Study Group #1, 2019 and Study Group #1, 2016). The lessons outlined in these books were designed to be studied, tested, and experienced by individuals who are seeking a better understanding of their life situation and the cause of personal conflicts, trials, and suffering that too often seem to occur for no apparent reason. Application of the principles outlined in the books can lead a seeking soul to a restored relationship with God. The bottom line of the philosophy of the readings is that the vast majority of our major difficulties in life arise from a single root cause, soul rebellion against God. Rebellion is a form of selfishness in which the soul thinks it knows better than God what is best for it or the soul becomes attached to a mental desire that is in conflict with God's purpose and plan for souls. The twenty-four lessons of the Search for God books outline a practical application of the two commandments of Jesus and the spiritual benefit to be gained by integrating their practice into everyday life. Neither the two commandments nor the Search for God readings promote the idea that belief in the historical figure of Jesus and repentance of sins grants the sinner immediate salvation and a ticket to a coveted place in heaven. This is a religious shortcut devised by man for personal comfort or to control other men when claims are made that salvation can only be granted through special representatives of God on the earth.

> *This [Gnosticism] is a parallel [to Christianity], and was the commonly accepted one until there began to be set rules in which there were the attempts to take short cuts. And there are none in Christianity! [Edgar Cayce reading 5749-14]*

Can the readings also be distorted and misused for evil purposes. Absolutely! The message of the readings can be corrupted as long as there are self-centered men or women who see a way to use them to gain advantage or power over others or who try to force others to accept their interpretation of the meaning of life and the nature of God and man. Man seems to have an infinite capacity to create evil out of the goodness that is God. The essence of the teachings of Jesus as understood and written down by the apostles and disciples of Jesus evolved from various manuscripts into a generally accepted canonical version of the New Testament of the Christian Bible over a period of some 300 years. The core of the New Testament is the two commandments of Jesus to love God with all our being and to love our neighbor as ourselves (Matthew 22:36–40). His first commandment is a reiteration of a commandment given to the Jews in Deuteronomy 6:5 many years before he entered the earth, and his second commandment is a restatement of the directive given to the Jews in Leviticus 19:18. Since then the Bible has been interpreted in many ways that insult the spiritual authority of Jesus by debasing and dishonoring his two commandments. Even in recent years the Bible has been used by various Christian denominations to justify the enslaving of millions of Africans and their descendants, by the Church of Jesus Christ of Latter-Day Saints to enforce the barring of persons of African descent from positions of authority in the church, and by conservative Christians to rationalize racism and white supremacy. The strength of the New Testament for future generations resides in the two commandments of Jesus. God has been trying to get humanity to put these commandments into practice for a long time. There is nothing else that is spiritually as important. There is no other belief or activity that God requires of us and there is no other topic in scripture that is as important. There is really nothing else that matters.

The equivalent core material in the readings, the essential message that contains everything that we need to know about God, our relationship with him, and our responsibility to him as souls created in his image is embodied in the 262-series readings that were used to compile the Search for God study lessons. These books present spiritual lessons that are a written version of the Way, the proper approach to life as demonstrated by the principled and moral activity of Jesus. They encapsulate the manner in which Jesus lived his life, the reason he lived his life in that manner, and the manner in which we must lead our lives to restore our awareness of God consciousness. They tell how our body and soul are connected, how the activity of one influences the other, and why it is necessary for souls to incarnate. The remaining readings mostly describe how individuals could, did, or did not apply the lessons of the readings to address specific health issues or to help them understand their present life situation and circumstances in relation to their previous application or misapplication of spiritual truth.

The search for religious truth leads many Christians (and members of other religious groups) into spiritual cul-de-sacs and dead ends that stunt the spiritual growth of their souls. The life of Jesus encapsulated and verified the truth that spiritual growth and perfection of the soul is predicated on our willingness to love God with all our heart, mind, strength, and body and love our neighbor as ourselves. This is the way that opens the mind to consciousness of the presence of God within and the realization and full awareness of our oneness and unity with God. In practical application it means we must manifest the biblical fruits of the spirit in every aspect of our lives. It allows our soul to understand and accept our role in bringing his kingdom into the earth and into any realm of consciousness that we may experience. There is nothing in his message that tells us which creed we must believe to get to heaven, informs us of the danger of going to hell if we speak of the preexistence of souls, or that gives us the correct number of candles to light or proper prayer to say during a particular religious ceremony. He didn't promote formalism and ritual except for baptism as a public display of commitment to a new way of life and the use of bread and wine as a method to bring to our consciousness the remembrance of his life on the earth, the sacrifices he made, and the ultimate purpose for his life and death.

Jesus didn't give the two commandments to Christians; he gave them to the world. These principles can be followed by anyone to the same spiritual benefit, whether or not they associate themselves with Jesus or Christianity. These are the truths that set men free of the bondage of materialism and selfishness that bind the soul to repeated cycles of rebirth and that expand the soul mind and draw the soul into a closer communion with God. The soul must be perfected before it can return to full companionship with God. It is up to each individual soul to make the conscious decision to initiate this process and engage in the appropriate thought patterns and actions that will ensure its success. This depends entirely on the soul's willingness to let go of all forms of selfishness and all manner of rebellion against God in spirit, in mind, or in flesh. We don't have to attend church to be aware that this better moral behavior will be beneficial to ourselves, our families, and the society in which we live, but we may well start attending church as we realize the transformational affect that such behavior has on the soul, the awareness that it brings of our connection to and relationship with God, and the desire to share our new awareness with like-minded individuals.

**The process of soul perfection can be
as long as the soul wants it to be or as short as God wants it to be.**

Are Religion And Science Compatible?

Science is the pursuit of the truth about the physical world around us through the process of observation, measurement, and analysis of physical phenomena and repeated testing of ideas and theories. Testing is an essential component of the process because it is a self-correcting mechanism that eventually detects errors in experimental procedures, assumptions, and theories or limitations that may only become apparent as experimental techniques are improved and extend the range and accuracy of measurements. This process leads to a comprehensive understanding of the nature of the physical world and the physical, chemical, and biological processes that keep it in motion. But no amount of scientific knowledge, no matter how well a theory predicts the behavior of a certain class of physical phenomena or how well scientists can isolate and map a specialized region of the brain and its particular functions, can lead to greater spiritual understanding. Scientists are often in awe of the majesty of the cosmos or the intricacy of a biological process, but they are trained to study these physical phenomena rationally. If every new discovery is viewed from a purely analytical reasoning perspective, it will forever remain as scientific knowledge and never will be transformed into spiritual understanding.

According to the readings, virtue is a soul quality that is expressed when we hold true to our higher spiritual ideal, the highest standard of righteousness and morality we can perceive and aspire to meet. When we strive to express this high quality of character in our lives, it allows us to begin to see the physical world from the perspective of a soul that is seeking a closer relationship with God. It can allow scientists to understand that their discoveries bring to light the intricate workings of God as revealed in the physical world of matter and energy. Although science can reveal the grandeur of God's creation, God the loving Father cannot be not found by using ever more sophisticated electronic instruments to probe and measure the distant expanse of space, the microscopic world of the cell, or the submicroscopic world where the electromagnetic, strong, and weak forces reign. Scientific knowledge can be the basis for a better understanding of God the Creator, but the awareness of God as Father resides in the soul mind, not in the knowledge of the physical world. When we hold science as the whole truth and exclude religious thought or hold religious thought as the whole truth and exclude science, we only see one half of the totality that is God.

Far too many Christians see science and religion as antithetical and try to use the Bible to disprove scientific theories and experimental results that don't conform to their particular interpretation of the Bible. Usually these efforts center around questions concerning the origin of the universe and the rise of biological life on the earth. How do the latest scientific theories on the creation of the universe, and the experimental observations that support them, mesh with the readings? The readings may not agree with current scientific understanding on every issue, but they do make it clear that science and religion are not only compatible, they both are seeking to find and understand God, even if that does not seem to be the case to many scientists and religious leaders. But the caveat to that assertion is that there must be a common desire and purpose to seek truth and a willingness to explore alternative explanations when the current theories, interpretations, or suppositions face inconsistent facts.

Science and religion are one when their purposes are one.
[Edgar Cayce reading 5023-2]

Science advances in a series of logical steps codified in the scientific method. In their initial formulations, theories often act more as a working hypothesis, or even speculation, rather than factual statements of universal truth. New ideas are gained during the direct observation of natural phenomena and the organization and study of previously collected data. These ideas are formulated into a conceptual or mathematical theory in an attempt to organize the ideas into a new physical law or extend a previously formulated law. The mathematical framework of the theory is studied to predict unobserved or undetected physical phenomena that may be amenable to experimental verification. Depending upon the outcome of the experiments, the theory may be modified or discarded. In either case, the theory and experimental results are published and made available to the wider scientific community so that others will be able to assess the quality and accuracy of the study and can attempt to reproduce the experimental results or devise and conduct new experimental measurements of their own design. Only theories that pass this rigorous testing and verification procedure ultimately survive to become part of the accepted scientific body of knowledge.

> *Science is organized common sense, where many a beautiful theory was killed by an ugly fact. - Thomas Huxley*

Scientists will iterate toward a truthful description of the physical universe. Every observation that science cannot explain or inadequately explains is not a source of shame or reason to manufacture a set of false facts to pretend there is no problem with the current state of knowledge. It is just another challenge to be met and another opportunity to rethink and refine our understanding of physical laws. Religious leaders, unlike scientists, generally don't believe they need to find truth through research and experimentation or need to question doctrines and creeds, because they are already certain they have the truth. It doesn't matter that each religion, even each denomination, has a different scripture or adopts a different belief system, publishes different doctrinal manuals, and insists on the use of their special mode of worship involving different rituals and songs that are designed for the faithful. Every one of these unique religious schools of thought was founded with absolute certainty that their truth is the real truth, a truth that they systematize within the context of scriptures, doctrines, bylaws, and rituals and usually require their members to believe and follow.

Religion has no formal rational method to establish truth in a manner equivalent to the scientific method. In the early Christian Catholic Church councils were convened and delegates spent days or weeks discussing doctrinal questions and assessing other matters of theology and passed judgment on what could be believed and what could not be believed if one was to remain within the newly organized and increasingly powerful church. During the period from 100 AD to 500 AD, there were several competing interpretations of the relationship between man and God and the words written by the early apostles, but these interpretations and the people who adhered to them were routinely pushed to the sidelines through excommunication and sometimes outright extermination of their leaders and followers. The truism that power tends to corrupt and absolute power corrupts absolutely is just as valid in religion as in politics. The founders of the Association for Research and Enlightenment were warned by the source of the readings about the danger of assuming they had ultimate truth and believing they had the right to force it onto others (254-87). They were told to get the research before the enlightenment, to test the principles set forth in the readings by applying them in their personal lives to confirm that they were a source for good and deserved to become a standard of behavior that should be offered to others. This method of assessing the truth contained within a revelation, prophesy, or scripture requires that the seeker set aside all selfish interests, personal ambitions and adulation of any leader to arrive at an unbiased and untainted assessment of the offered truth.

There are many instances herein where literal biblical interpretations and material from the Edgar Cayce readings are compared to current scientific theories and knowledge in relation to physics, anthropology, and biology. Biblical "science" is demonstrated to be false when compared to established scientific facts and theories on cosmology and evolution. Many of the readings used herein mention topics and ideas that are not compatible with science or some religions (concepts such as reincarnation, astrology, ancient civilizations, the nature of God and soul, soul activity on the earth, the purpose for the universe, and the nature of electricity). Many of these ideas are unverified and currently unverifiable by science. They are not declared incorrect or wrong because of that unless they conflict with an established scientific principle that has proven them false. They are used herein to compare, where possible, information given in the readings to the current scientific understanding about some aspects of the physical world. They should not be declared scientifically right or true just because they came from the readings and neither should biblical passages be declared scientifically correct simply because they came from the Bible. Someday they may be scientifically proven correct or incorrect. Either way will lead to advancement of knowledge and a closer approach to Truth. Some of the ideas presented in the readings may remain in the realm of philosophy and religious thought until they can be assessed from outside the physical universe.

Observations in the physical universe by the human body are primarily by the exchange of energy via photons (sight), molecular vibration (touch and hearing), and ions in solution and molecular structure (taste and smell). But the aspect of God that relates to us personally is experienced in the mind of the soul and is not recognized by using the five senses or by using electronic instruments that mimic and enhance the senses. There is a problem with looking for God in the physical universe if we expect to find evidence of a spiritual being of higher order than man, but there is no problem with studying the universe for evidence of God's creative activity. Scientific theories and explanations of physical, chemical, and biological phenomena are not antithetical to or incompatible with the existence of God. Scientific interpretations based on observed and measured natural phenomena may be incorrect or incomplete in the early stages of their formulation but usually become more correct over time as results are verified, analyses are refined, or conclusions and experiments are validated or demonstrated to be false by peer scientists. Initial interpretations of physical processes may be limited or bounded in scope and subject to future modification because current technology is not capable of making measurements to the accuracy required to move science beyond the present limitations.

The facts and truths that evolve out of scientific exploration help explain what God has accomplished with his creative activities as they pertain to matter and energy and may give insight into how God thinks and acts. This new knowledge doesn't and shouldn't take the place of religion, but neither should religious faith or belief be a reason to ignore or ridicule the results of scientific investigations, especially those that have yielded results that have withstood the test of time during the era when the scientific method has been used to test and refine theories by making repeatable and verifiable observations and measurements of their predictions. Religion should not look to science for confirmation of God's existence and science should not reject religion because God or spirit has not been observed or detected by its instruments. Religions should not claim their scriptures are the source of scientific truths, just as no branch of science need claim to be a new religion.

Science seeks to explain the universe created by God.
Religion seeks to explain God who created the universe.

The universe is not a place created by God separate from and independent of himself but is an expression of God arising out of himself in a manner designed to create the conditions necessary for the formation of atomic matter and the spacetime structure that holds it all together. Science studies God just as religion studies God but approaches this study from within the manifested universe of matter and energy, while religion takes a more difficult approach toward the same from outside the material universe through the activity of the mind and soul. Science and religion are like the blind men and the elephant, each giving a different description of the same being because they are touching a different part of a single organism. Just like the blind men, science and religion are at odds because they engage the universe from different perspectives, and their adherents are convinced that they are right and that the other is wrong.

However, science and religion are fully compatible. They incorporate a different approach in man's attempt to discover the truth about himself and his origins and the nature of the universe in which he lives. They both are trying to find truth but are seeking it from a different perspective and with a different purpose and goal in mind. Science, with its emphasis on logical thinking and mathematical rigor, application of the scientific method of theory formulation and prediction followed by experimentation and observation, and emphasis on peer review of conclusions is self-correcting in the long term. This is more apparent and perhaps truer of the so-called "hard" sciences that study the physical laws of the universe using experimental apparatus to measure physical properties and matter-matter or matter-energy interactions than the so-called "soft" sciences that study social or psychological phenomena that are less amenable to direct observation or investigation. Advances in physical science are often made in conjunction with the increased sensitivity of newer experimental equipment. A theory considered valid for many years and found to be correct within the limitations set by the sensitivity of available instruments and measurement techniques may fail to correctly predict the results of an observation made at higher precision with more sensitive instruments or using a wider spectrum of electromagnetic radiation. This does not mean that the theory was wrong, only that it described a limited portion of a larger reality and that advances in instrumentation have pushed the range of observations beyond the range of application of the original theory. Perhaps the practice of meditation and consistent application of the principles of behavior outlined in the two commandments of Jesus can have an effect similar to an advance in scientific instrumentation and allow mankind to better perceive and understand its relationship with God and its own spiritual nature and heritage.

Science cannot (currently) detect or prove the existence of God or the spiritual realm, and scientists generally have not concluded that such investigation is warranted. Whether it will acquire the necessary tools in the future is an open question to many, but the readings are clear that the potential for scientific study and measurement of the mind and spirit is real. At the moment, most scientists don't want to believe in or risk investigation of spirit and most religious leaders don't think that science can or should have anything to do with matters of spirit. As science expands our knowledge of cosmology and molecular biology, we come closer to understanding the mechanics of the physical universe and the biological life it supports but that, in and of itself, does not bring us closer to understanding the meaning of the universe or life. Major discoveries in science usually follow the discovery of new and more accurate methods of observing the physical world, but sometimes advances are the result of looking at existing data in a new way that resolves a paradox.

Refinements in observational data and reassessment of old data in light of new knowledge can expose shortcomings and limitations in existing scientific explanations that were not previously known and lead scientists to modify an existing theory or formulate a radically new theory. Albert Einstein was brilliantly and famously successful at devising new concepts of reality to solve seemingly intractable scientific problems. His new theories often postulated or predicted previously unobserved properties of the natural world and set in motion a flurry of activity to develop new experiments or observational methods to ascertain the veracity and accuracy of his theories. Scientific studies of the brain have moved beyond trying to understand the mechanics and role of neurons, and scientists are beginning to explore the origins of the mind and consciousness as part of the brain's neural activity. Perhaps this is the first step that will lead scientists to understand the mind in a way that will transcend the physical.

Will psychic phenomena ever, or within say 50 years, be accepted and provable on directly scientific measurements; that is on meters (instruments) and mathematics? If not,

> *why? (A) When there is the same interest or study given to things or phases of mental and spiritual phenomena as has been and is given to the materialized or material phenomena, then it will become just as practical, as measurable, as meter-able, as any other phase of human experience. [Edgar Cayce reading 2012-1]*

Scientific discoveries are a remarkable testament to the power of mind granted to souls. The ability to reason logically from observed phenomenon, to develop abstract theories, and to use those theories to predict, test, and validate new ideas about the behavior of matter has allowed man to transform his environment from the natural world to the man-made world. Scientists today can design advanced machines to measure particle properties through subatomic particle collisions and matter-energy interactions or observe the world-shattering collisions of two galaxies. However, when we become enamored with the power and wonder of the mind and the machines we can build and lose sight of the spiritual component of our being, we create a mental barrier that prevents us from opening ourselves to the Spirit that resides deeper within our minds. When our scientific achievements and knowledge are used to support the notion that the only reality is the physical world and that God does not exist because he cannot be seen or measured, we block any possibility of learning about the spiritual aspect of our being. We prevent God from entering into our hearts, and we create obstacles that prevent others from knowing God when we try to convince them that science is a replacement for God.

According to the readings, the spiritual world operates on a higher level of energy and vibrational state and exists in a different dimensionality than the material world. All observations that man makes of his physical environment are defined by the range and sensitivity of the sense organs in the body and mechanical and electrical equipment that simulates and extends the range of these senses. Perhaps scientific instruments will one day be constructed with the capability of measuring phenomena operating at vibrational rates higher than those currently observable (perhaps such as electromagnetic radiation of higher frequency than gamma rays) and science will recognize that it has pushed into the realm of spirit. Then again, even if the spiritual realm is pierced by scientific instruments, scientists may only assume they have peered a bit deeper into the physical realm and will be unable to recognize that they are making measurements and indirect observations at the intersection of the material and spiritual worlds. The perspective that comes from looking at the universe from the inside instead of looking at it from the outside causes many scientists and nonscientists alike to conclude that there is nothing but nothingness beyond the boundaries of the known physical universe.

> *The idea that the Laws of Physics can vary throughout the universe is as meaningless as the idea that there can be more than one universe. The universe is all there is; it may be the one noun in the English language that logically should have no plural. The laws governing the universe as a whole cannot change. What laws would govern those changes? Are they not also part of the Laws of Physics? - Leonard Susskind*

Will science discover the spiritual realm as it looks deeper and more intently into the underlying foundation of the physical realm? Will the findings of science enlighten humankind to the possibility of a divine hand at work in the cosmos or detect evidence of his presence in the natural world? Many Christian nonscientists and many Christian scientists would say that we already see that evidence every day. Will science someday prove the existence of God? All of these might well be true, but it doesn't say everyone or even most people would recognize the significance of such a discovery even if it were made. There would still have to be recognition that the discovery uncovered something outside of the realm of the physical world or that the physical world is an extension and expression of God and that the feature or material or force being studied is really a form of spirit and not a previously unknown extension of materiality. Maybe there is no sharp boundary that separates material from spirit, and instead there is a gentle transition between the two that makes it difficult to determine whether one is inside of the universe or outside of the universe. How does one realize a new discovery is something outside of the known universe when it is made by individuals and instruments from inside the universe? Despite the limitations imposed on us, the indirect methods we have used to observe and study the creation of the universe and the life that inhabits one of its planets has been remarkably fruitful over the past 150 years. But will we ever be able to fully understand the initial creative expression that set the universe into motion by peering at it from the inside, as opposed to viewing it from the outside?

What Or Who Is God?

This makes for that which is in keeping with a fact that has been from the foundations of the earth - God IS! and they that would know Him must believe that He is and may be made manifest in their dealings with their fellow man! [Edgar Cayce reading 254-101]

What is the first cause? That which has brought, is bringing, all life into being; or animation, or force, or power, or movement, or consciousness, as to either the material plane, the mental plane, the spiritual plane. Hence it is the force that is called Lord, God, Jehovah, Yah, Ohum [Ohm?], Abba and the like. Hence the activity that is seen of any element in the material plane is a manifestation of that first cause. One Force. [Edgar Cayce reading 254-67]

Reading 254-101 is one of many that make it abundantly clear that God exists and also makes it clear that God is not to be imagined as something or someone that is visible to the naked eye if we are lucky enough to be in the right place at the right time with the right desire and mental attitude. He is not to be searched for and found in some remote location on the earth or even somewhere in the outer reaches of the cosmos although he is present in all of those locations. That is not to say that we cannot find evidence of him in the physical world or that we should not marvel at the evidence of his manifestations in the world, but seeking physical evidence of God should not be the purpose that gives us hope, inspires us to be better, and drives our desire to find him. The opportunity to see God "in person" or observe him in the physical world is not a reason that should cause us to search for God or define the manner in which we should look for him. The readings specify two important criteria that are necessary for a seeker to discover God. They are 1) to believe that he exists even though we are unable to see or sense him (faith), and 2) to manifest his presence through the way in which we treat our fellow man (loving service).

Faith, hope, and love are all activities of the soul. They don't have their roots in the physical body or the material universe and don't leave a physical trace from disturbance of atomic matter, so are not readily amenable to scientific study. They originate in a subconscious mind that desires to express those feelings. Faith is an attribute of the soul that can open the conscious mind to the sure knowledge that spiritual forces can and will flow from God through the soul and become manifested in the physical world. Faith is not derived from physical experience, but faith can be strengthened by observing its effect in the world. The exercise of faith increases faith, but faith does not arise from any expectation by God that we blindly follow a particular religious dogma or creed, no matter how large or powerful the church or how persuasive or charismatic the pastor or priest that promotes the doctrine. Faith is the ability to recognize and become consciously aware that there is a higher force or power that is greater than anything we observe in our daily activities, something out there of more value and importance than anything in this world that our conscious mind can desire or crave.

So why should we be good to other people? Is it because of the good feeling we usually get in return for our effort? Do we think we are racking up a higher score with God with each action of goodwill toward other less-fortunate individuals? Do we worry that our goodness grade will not meet the standards that God has set for us? If any of these apply, we are doing the right things for the wrong reasons. Our actions may be materially beneficial to other individuals, but there is little spiritual benefit or value to ourselves. Why? Because God does not ask us to perform these good works only to relieve the suffering of our neighbors, although that is a noble goal of any service activity. When we do good works in a consistent and purposeful manner, we actually alter the mental pattern and spiritual awareness of our soul mind and open ourselves to let God work through us for his purposes rather than us working for him in ways that, in our arrogance, we deem to be better or more appropriate. According to the readings, this is the mindset that got Adam into trouble.

During our lives we are presented with opportunities to demonstrate spiritual truths to our friends, neighbors, and even our enemies. If we lack faith, we lack the ability to recognize the spiritual dimension of these situations and implement the necessary and correct actions that lead to spiritual growth. Faith keeps us attuned to the spirit within us and opens the channel of communication and communion with God that is needed if we are to be channels of blessings to others. Without faith in the existence and presence of God, we tend to remain focused on ourselves and the material world, and that tunnel vision causes us to ignore promptings and guidance from the Spirit within and make life decisions based solely on material and selfish interests. For example, our selfish interest in acquiring money for comfort or security can override a spiritual prompting to share those resources with those that are less fortunate.

> *These experiences in the earth are not that one may glory in self, whether by fame or fortune or a wonderful accomplishment to the detriment of other souls, but that each shall ever be a channel through which blessings and help may come to others. [Edgar Cayce reading 2509-2]*

The apostle Thomas is often brought to task in sermons because he did not have faith that Jesus had resurrected despite the previous promise made by Jesus and the testimony of other apostles (John 20:25). Because of his lack of faith, Thomas needed evidence in the form of material proof that could only come through the physical senses. It took the appearance of Jesus in the upper room in the presence of Thomas before he was fully able to accept that Jesus had indeed resurrected. This need to see physical evidence to build belief or confidence is necessary for those focused on the material world. It is a good, necessary, and valid attitude for a scientist but can impede spiritual advancement for those who reject the existence of God or the spiritual realm because they cannot be seen, touched, or heard. Actually, Thomas is treated unfairly by Christians today because he wasn't alone among the apostles in his disbelief of the resurrection of Jesus (Mark 16:9-14, Luke 24:9-11).

The apostle Peter had faith, but it was a newfound faith that was not yet strong. His first steps on the water when Jesus invited him to walk from the boat to the shore showed this budding faith, but as his mind began to doubt what he saw with his eyes, he sank into the water (Matthew 14-28-31). The soul mind is powerful, but the physical conscious mind too readily focuses on sensory information from the material world to the exclusion of guidance from the spiritual world. When the conscious mind dominates, the soul awareness of the spirit can be drowned in the noise and dissonance of material sensations. It takes an element of faith and sustained effort to recognize the existence of the spiritual world and discern the power and ability of spiritual forces to act on and in the material world, especially when the manifestation of that power is counter to our understanding of everyday experience and physical laws. We must accept that God IS, that he exists, before we can have any relationship with him. Otherwise we close ourselves to the awareness of his presence and shut off any opportunity to personally interact with him.

> *Is it correct when praying to think of God as impersonal force or energy, everywhere present; or as an intelligent listening mind, which is aware of every individual on earth and who intimately knows everyone's needs and how to meet them? (A) Both! For He is also the energies in the finite moving in material manifestation. He is also the Infinite, with the awareness. And thus as ye attune thy own consciousness, thy own awareness, the unfoldment of the presence within*

beareth witness with the presence without. [Edgar Cayce reading 1158-14]

Reading 1158-14 alludes to two ways in which humans perceive and interact with God. We experience and know God impersonally through our senses because the physical universe is a manifestation of his Spirit. Believers, nonbelievers, and agnostics alike in every location and at every time in human history have experienced God through materiality. However, living in a manifested expression of Spirit in a material body does not in itself impart an awareness of God to the mind of man. Our mind does not typically think of physical objects as an aspect of God. Perhaps we are most likely to make the mental connection between God and materiality and understand the universe as an expression of God when we are standing on the rim of the Grand Canyon, are watching the geysers of Yellowstone, or are observing similar magnificent examples of nature on a grand scale. We experience and know God personally through our mind and emotions as we willingly and actively seek him and desire to have a personal relationship with him. The reading reiterates that the criteria for becoming mentally aware of the presence of God is by having faith that he exists and by exercising that faith in a practical affirmation of his presence through loving service toward our fellow man. This outpouring of divine love is a manifestation of an inherent God-given attribute of the soul that always seeks to be expressed, always desires to flow from our soul into the world. It strengthens our conscious awareness of our relationship with God. It strengthens our faith.

God has been described by different religions using many different words to identify his perceived qualities and characteristics and attempts to describe the ways in which he reveals himself to us. Some of these descriptive terms attempt to explain his grandeur, his emotive state, his relationship to us, his perceived oneness, his longevity, his inexplicability, and his unfathomable mystery. Christian and Jewish descriptions of God are given in the Christian Bible. The Bible tells us that God created the heavens and the earth (Genesis 1:1), is all powerful over things heavenly and earthly (omnipotent; Jeremiah 32:17 and Revelation 19:6), and is limitless in his knowledge and wisdom (omniscient; Psalm 139:1–6 and Romans 11:33). God transcends the known universe (Psalm 113:4–5), is just in his dealings with men and impartial in all of his judgments (Psalm 75:1–7), and is always present everywhere (Psalm 139:7–12). God is one, and there is no other God but him (Deuteronomy 6:4 and 1 Corinthians 8:4–6). God is faithful in that he keeps his promises and will always honor his covenant with man (Deuteronomy 7:9). God is described as good (Psalm 119:65–72, Mark 10:18 and Luke 18:18–19) and loving (1 John 4: 7–10). God is graceful and freely offers his grace to us even though we are undeserving of it (Ephesians 1:5–8), and God is Father to us, our creator, and will be our comforter, but only if we will seek to be his children (2 Corinthians 1:3–4).

Islam teaches that God has ninety-nine names that describe his attributes or characteristics. The list of these names vary slightly from one source to another but are all extracted from the Qur'an and hadiths, a collection of traditions and sayings of the prophet Muhammad. The specific forms of the listed names are reserved only for reference to God but are used in modified form as given names for men and women. The first six of the ninety-nine names of God are Ar-Rahman the All-Compassionate, Ar-Rahim the All-Merciful, Al-Malik the Absolute Ruler, Al-Quddus the Pure One, As-Salam the Source of Peace, and Al-Mu'min the Inspirer of Faith.

The Sikh religion accepts many names for God, including names given by other religions. Among the hundreds of names listed in the Guru Granth Sahib Ji that describe and define the qualities of God are All is One (Ik Oankar), Truth is the Name (Sat Naam), Primal Creator (Karta Purakh), Without Fear (Nirbhau), Without Enmity (Nirvair), Timeless (Akaal Moorat), Unborn (Ajauni), Self-Existent (Saibhang), Grace (GurParsad), Truth before Time (Aad Sach), Truth throughout Time (Jugaad Sach), Truth Here and Now (Haibhi Sach), and Truth Evermore (Hosibhi Sach).

The readings also present several descriptions of God that give a sense of his character and his relationship to his spiritual and physical creations. They include qualities or attributes such as God is One, God is Spirit, God is Love, God is Law, God is Life, God is Truth, God is First Cause, and God is Creative Force. They describe his nature and activities as God is eternal, God is not a respecter of persons, God is a consuming fire that will purge everyone, and God is an all-wise, all-inclusive, all-manifesting force in the experience of man. God is a fact; God is the avenger; God is a living God; God is knowledge, God is merciful, just, and patient; God is longsuffering; God is mindful of his children; God is ever present; God is the same yesterday, today and forever; and God is a zealous God, desiring that all souls should be one with him and work to bring the knowledge of him to mankind. Six of these attributes are particularly relevant to the following text on the creation of souls, the universe in which they incarnate, and the purpose of human life and will be discussed in the paragraphs below. The readings disagree with the Psalmist who declared that God knows our every thought and action (Psalm 139:1–6). The one thing that God does not know is what we will do with his gift of free will and independent thought. Will we use it for self-gratification and self-indulgence, or will we use it to seek a closer relationship with our creator?

> *God Himself knows not what man will destine to do with himself, else would He have repented that He had made man? [Gen.6:6] He has given man free-will. MAN destines the body! [Edgar Cayce reading 262-86]*

> *God Himself knows not what a man may do with his own will from day to day! [Edgar Cayce reading 311-9]*

> *Having given free will, then, - though having the foreknowledge, though being omnipotent and omnipresent, - it is only when the soul that is a portion of God CHOOSES that God knows the end thereof. [Edgar Cayce reading 5749-14]*

God is One

> God as One cannot be fully comprehended by the conscious mind but is revealed as man learns his true nature as an incarnated soul and as his soul reawakens to its true relationship with God.

None is convinced in that science or religious convictions are one. The first lesson for SIX MONTHS should be ONE - One - One - ONE; Oneness of God, oneness of man's relation, oneness of force, oneness of time, oneness of purpose, ONENESS in every effort - Oneness - Oneness! [Edgar Cayce reading 900-429]

It was he who gave some to be apostles, some to be prophets, some to be evangelists, and some to be pastors and teachers, to prepare God's people for works of service, so that the body of Christ may be built up until we all reach unity in the faith and in the knowledge of the Son of God and become mature, attaining to the whole measure of the fullness of Christ. (Ephesians 4:11-13)

The Lord thy God is One (Deuteronomy 6:4 and Mark 12:29) is a fundamental tenet of Christianity, Islam, and Judaism. It is commonly interpreted to mean that God is single, solitary, unique, and individual as opposed to pagan concepts that admit the existence of a multiplicity of gods. The teaching of one god set the Jewish people of 2,000 years ago apart from their neighbors and was part of what made their religion unique. But to understand this expression by only this narrow interpretation is to limit its scope and obscure the larger meaning it carries. According to the readings, the phrase "the Lord thy God is One" also conveys the broader concept that God is One with his creations, whether that be the created souls that have the potential to become one in mind with God, the created material universe in which they have the opportunity to achieve that potential, or the created body that allows souls the ability to express their concept of God in physical lives and perceive the consequences of failure to choose God over self.

God is the one Spirit of which everything is composed. All material and nonmaterial realms are an expression of God's will, and all physical objects are a form of His Spirit in motion and activity. God is the all-pervading Love that is limitless, that exceeds all bounds, and that is unconditional. God sought expression for that love by endowing a part of his creation with the ability to express and reciprocate his love. That creation was our souls, made in the image of God with the inherited quality of an independent will and mind and the potential to love him as fully and completely as he loves us. God is the Law that gives the souls of men the opportunity of activity in a realm of causality so that they might understand how their selfish thoughts and actions have caused them to lose awareness of their creator and can regain their true relationship with him. God is the one source of Life that transformed the earth from a typical sterile rocky planet to a wonderland of plants and animals from the lowly amoeba to the gigantic Tyrannosaurus Rex to the perfect human form through which souls can expresses their understanding of God. God is the Truth that transcends the accumulated knowledge, understanding, and wisdom that man has acquired through inductive and deductive reasoning or intuition.

> *O, that all men would know, "Know, O ye children of men, the Lord thy God is ONE!" Each spirit, each manifestation of LIFE is ONE, and a manifestation either in this, that or the other sphere, or scope, or space of development TOWARDS the knowledge, the understanding, the conception of that ONE - HIM - I AM - God - Jehovah - Yah [?] - ALL ONE! [Edgar Cayce reading 262-32]*

The idea of Oneness is central to the way in which the readings express our relationship with God and each other. As we attune our minds to God through physical actions that conform to the fruits of the spirit, we become closer to God in mind. To be attuned means to bring into a state of harmony or to resonate with a shared common purpose and ideal. To be attuned, or in tune, can mean to vibrate at the same frequency, thus the readings suggest that we strive to alter or raise the vibrational levels of our consciousness from the lower, more base frequencies associated with physical incarnation and the cares of the material world to the higher, more spiritual frequencies associated with a mind that communes with God. As our wills become more spiritual through intentional decisions that resonate with God's purpose for us, we are able to achieve a state of oneness with the will of God and become a channel through which his essence and purpose for us can shine forth and illuminate the path for others to follow. Jesus was well aware of this concept of oneness, which he expressed clearly when he stated that "I and my Father are One" (John 10:30) and "He that hath seen me hath seen the Father" (John 14:9). Jesus was not God (John 5:18), but he dared to make himself a true physical manifestation of the mental and moral attributes of God, that is, one with God in mind and spirit, and he encouraged us to set our sights beyond the cares of the physical world and walk the same path that he chose to take.

Jesus declared to the Jews in the Temple at Jerusalem that they were gods (John 10:23–34), by which he meant gods in potential and in the making and did not mean that they were the almighty God. He was only reminding them of their own scripture which, in reference to those officials who stood in judgment of others, stated in the Psalm of Asaph that they were gods (Psalm 82:1–8). All of us have the potential within our mind and soul to walk the path demonstrated by Jesus and thereby come to understand our oneness with God and our ability to share in his glory by glorifying him with our thoughts, words, and actions. He broke the trail for us so that our walk through life would be easier, and he promised to take upon himself part of burden we carry when we face seemingly overwhelming circumstances beyond our control. We still must push aside and overcome the obstacles we have created and continue to create for ourselves every time our minds entertain selfish thoughts, especially when we act on those thoughts to the detriment of ourselves or those whom we meet during our journey through life. We are not capable of being God, the part cannot become the whole, and the arrogant selfishness of such a desire would forgo the possibility of its occurrence. We are capable of being like God (257-201); that is, we can create by using our mind to direct spirit and manifest activity and we can express many of the attributes of God such as love, kindness, and patience.

God is not the master of some spiritual realm far off in the outer reaches of space who looks down upon a wayward humankind struggling to successfully navigate the trials and temptations of the physical realm, ready to bestow the honor of admission into his spiritual realm to those who make the greatest effort to physically and mentally overcome, or at least endure, the hardships they encounter. What we call the spiritual realm, mental realm, and physical realm are all equally expressions of God, and he is constantly present in all of them. They are one because God is One, and all that exists is an expression of God's creativity in action. The biblical passages and readings that stress the concept that the Lord thy God is One express more than the idea that there is only one deity; they also have the greater and deeper meaning that God is unitary. He is all that is, and there is nothing else. God is a single, indivisible, and all-encompassing being and force, having no equal. God is One, but more important, all that exists is a portion and expression of God.

Unity through cooperation is a form of oneness. Unity comes from acceptance of others and respect for others despite our superficial human differences. God desires unity among his people because unity of purpose under God can bring people together in a way that allows them to be a stronger power for good in the world than they can be as individuals. It is not so much that we need to unify ourselves through some collective physical action, but that we need to recognize that our souls should be unified on a mental and spiritual level because we each were created as a portion of the life force that manifests through us and that is the expression of God within us. There is a peace that comes with unified thought and action based upon a desire to seek the will of God. Disunity among nations and states leads people into tribal or national divisions that ultimately cause disharmony, conflict, and war. Unity in the Oneness of God is the true and certain unifying force of humanity, not the temporary and false illusion of unity provided by strong or autocratic national leaders or that is secured by the military campaigns of great generals. The unity of oneness in God is so strong that it brings people together with the same strength of purpose that existed between Jesus and God, yet also allows them to maintain their individual natures and independence.

The maxim that a house divided cannot stand holds true for a nation and also holds true for God's people. Spiritual unity does not mean that everyone will become the same, having the same desires, thoughts and beliefs, and engage in the same activities. Spiritual unity among Christian denominations will be achieved when people having different religious beliefs put aside any self-centered motivation or self-righteous conviction that they have special knowledge of the only and true way to God and make a contribution toward the spiritual and material improvement of others in the name of Jesus the Christ. Unselfish loving actions and service toward others encourages societal unity and fosters a sense of community that mirrors divine oneness. Love is the glue that binds it all together and that will cause unity to materialize and spread through the diverse world population. Cooperation is more than working smoothly with others to accomplish a difficult or complex task. It has a spiritual dimension which is realized by opening the mind and heart to the presence of God and allowing him to work through each of us in service to others. This is true cooperation that leads to unity in purpose and fulfillment of God's plan for humanity and souls.

> *Though there may be many approaches, cooperation in the activities – as in the universe – brings the harmony of the universal activity; as does cooperation in human experience bring harmony and peace; while egotism and self-assertion and self-exaltation and self-indulgence bring inharmonious experiences, and the activity of turmoils, wars, strifes. [Edgar Cayce reading 1297-1]*

At what point can personal differences be seemingly insurmountable and threaten unity and oneness in God's purpose? Does the existence of multiple religions and religious denominations violate the concept of spiritual unity and oneness, or are different religious belief systems a necessary intermediate stage in human and soul spiritual evolution as each individual strives and struggles to understand the Truth that is God? Perhaps different belief systems allow those who are exploring an emerging interest in their spiritual nature to learn about God in ways that are more appropriate to their level of spiritual maturity or more necessary or palatable for the illumination of their soul mind. The challenge is to make sure that different beliefs do not lead to division, disunity, and conflict among religions and religious denominations.

God is Spirit

God as Spirit is the underlying fundamental inspiration and motivation that drives all creative activity and is the source of all manifestations of life in spiritual, mental and physical realms.

Spirit is the First Cause, the primary beginning, the motivative influence, - as God is Spirit. [Edgar Cayce reading 262-123]

God is the spiritual being that existed long before the physical universe was imagined or conceived and is the Creative Force that launched the material universe and formulated the physical laws that constrain the motion and interaction of matter and energy. The physical consciousness, the seat of human reason and analysis, has difficulty comprehending a nonphysical being and a nonphysical realm. In his own words, given to Moses as a way to describe the unimaginable to the finite and world-bound minds of the Israelites, he called himself the "I AM THAT I AM" (Exodus 3:14). In other words, there are no adequate words or concepts in human languages that are able to fully communicate the unbelievable majesty and other-worldly grandeur of the Creator. He is, he exists, and all attempts to describe him in terms understandable to the human physical consciousness are inadequate.

God is infinite and timeless. He is infinite not because his spiritual equivalent of physical spatial dimensions extends infinitely distant in all directions but because he encompasses all that exists and because everything that exists is, in some way or another, an expression and part of God. He is timeless not because he has no beginning and will have no ending in the physical sense of time but because time has no meaning for him and is not a property of his spiritual realm. The entire concept of time as we understand it from our physical experiences has no meaning in the realm of God, and he is not bound by the constraints of a spiritual equivalent of time. Time as an ordered sequence of events is not part of the structure of his spiritual realm and is not part of his consciousness or the consciousness of souls, and events in his realm are not causally connected in the manner perceived by the physical consciousness.

> *"God is SPIRIT and seeks such to worship him", or be one WITH Him, knowing self then to be self - yet one with that Spirit, making all things possible through the application of that understanding of Spirit in life's experiences or life's conditions, in whatever form they may be presented. [Edgar Cayce reading 106-18]*

His essence is threefold: Spirit, the etheric substance of his being that exists in a dimension or dimensions higher than, or other than, the material world we inhabit; Mind, the force he uses to direct and mold spirit to his creative glory and; Will, the intelligence by which he makes choices and guides Mind to the completion and fulfillment of his purposes. Spirit is the substance of God that is the foundation of all that exists in all realms and all dimensions. To call spirit a material would cause confusion to the conscious mind because we naturally associate material with matter, a compact form of condensed energy. Spirit is much more akin to energy than to matter. Spirit as a medium or spiritual substance is a form of energy of higher vibration (900-89 and 4735-1) than any energy we humans presently can experience through our sensory organs or can currently measure with our most advanced scientific instruments. It is energy of a higher vibrational state than the most energetic gamma radiation that has been detected in the physical universe. Spirit is not the substance of which God is made or a substance that exists outside of God that he uses or manipulates at his convenience. Spirit is the substance of which God IS, the essence of God himself. Yet, spirit is also the source substance from which the structure of all nonphysical forms and all physical matter emerges. Spirit is the motivating influence of all spiritual beings, all matter in the universe, and all earthly biological life; the precursor and life source of all created things.

> *God, the first cause, the first principle, the first movement, IS! That's the beginning! That is, that was, that ever shall be! [Edgar Cayce reading 262-52]*

> *What is the First Cause? That from which all emanates, the SPIRIT of the force or influence itself; breaking itself upon the atomic structures about same, bringing those influences as it associates itself one with another in its varied forms of atomic structure. [Edgar Cayce reading 2012-1]*

The readings describe Spirit as "First Cause" for good reason. There is no underlying substance of which spirit is composed. It is not an amalgam or composite of more basic components because it is the most fundamental and elemental ingredient of all realms, spiritual and physical. Spirit is God and God is Spirit. Spirit is the underlying cause, activating force, and motivating power of all that exists. It is the ultimate, complete, and total source of light, both in the sense of the electromagnetic spectrum of the physical universe and in the sense of a higher state of consciousness and understanding to which the mind of man can aspire. It is also the power that gives life to all biological organisms that exist or ever existed on the earth. Spirit is the unique and unadulterated substance that is the parent substance of all that exists in the spiritual and physical realms. How can we know and understand such a higher-order intelligence that is so otherworldly, so esoteric and exotic in comparison to our mundane earth-bound existence and apparently so far removed from our everyday life and environment? The soul shares spirit with its creator and can perceive the spirit of God and the God-conscious soul mind can commune with mind of God.

God is Love

> God as Love is satisfied in the creation of souls that have the eternal opportunity to bask in and absorb the warmth of their creator's love, to reflect and radiate that love as a beacon and testament to their creator, and to share and experience that love among themselves. Souls can share love with their creator because God created souls with the ability to fully express the divine attributes that make up the Spirit and essence of God. Souls have the potential to become true loving companions of God and to share the joy of his creative activity.

It is again as Infinite Love and Divine Love. Infinite Love is the love of God, while Love Divine is that manifested by those in their activities who are guided by love divine. These bring Happiness and the experiences of joy; not mere pleasure, not gratification of any of the material things. [Edgar Cayce reading 262-111]

God is an all-pervading spiritual manifestation of love. He is the epitome and essence of love, the most perfect expression of love in the spiritual realm and material realm. God not only expresses love for his creations, but is the standard and ideal against which all other expressions of love are to be measured. He is pure love, and his love is always pure. We owe our very existence, as souls and as human beings, to this most exalted attribute of God. For God, or anyone, to give love, there must not simply be an object to be loved, but the recipient must be a sentient being that can receive love into itself and that has the capacity to return that love. As a loving being, God desired companions who would bask in the warmth of his love, who would emotionally comprehend his love, and who would willingly and gladly reflect his love back to him. God created souls to fulfill this purpose, to be intelligent and loving companions who could share his penchant for creative activity.

Love cannot be forced, commanded, or coerced, so God can only offer his love to souls but cannot demand or require love from souls. He has to leave it up to each soul to make its own choice as to how it will respond to such an offer. Although God is not a person, our souls can choose to make that offered loving relationship as personal as we wish to make it. God is accessible to each of us through our soul mind and we can communicate directly with his spirit or presence if we so choose. God's expressions of love are pure and selfless. Human love is an imperfect and watered-down version of God's infinite love that souls have contaminated and diluted with selfishness. Souls project this flawed love onto the earth as thoughts that activate words and actions in human relationships that too often pervert the pure love of God. The lack of love in human interactions and the abuse and misuse of our fellow human beings for personal and selfish reasons is a visible symptom of the spiritual separation between souls and God. Each soul is responsible for its loss of awareness of God and inability to discern the love of God because of its selfish and self-centered thoughts and activities that dishonor God.

> *Each individual spirit [soul] then is only a portion of His Spirit. Not that God is separated. But in His love, in what we call infinite love, boundless, the unbounding grace and mercy and patience and love and long-suffering, these have brought to the Father the thought of the lack of, the wonderment of, companionship. [Edgar Cayce reading 262-115]*

What made God desire companionship? How long did he exist before he concluded that loving companionship was something he wanted to experience? What kind of stimulus or prompt did it take for God to realize that something essential was missing from his life, that there was nothing in all of his creations capable of reflecting his most pertinent and distinguishing quality? Were there no other creatures in any of his creations capable of giving and receiving love before he caused souls to come into existence to alleviate his loneliness? Did he reach a point where the ministrations, attentions, and servile behavior of angelic beings that could show loyalty and zeal but could not truly show love became unsatisfying and inadequate from an emotional aspect? What made him come to that realization? How long did it take for the idea to create a soul to incubate and come to fruition? Did it just never cross his mind to make such an audacious move until the thought hit him, "I am lonely" or "I wonder what it would be like to actually share love?" When we say God is eternal, we usually mean that he has no beginning and he will have no ending. Where does the creation of souls fit in the eternal timeless existence of God? Were souls created somewhere in the middle period or more toward the beginning, and does it even make sense to ask that question? Of course, no matter when this creation took place, God already existed for an infinite period before the event and will continue to exist for an infinite period after the event, but it is interesting to speculate about soul creation in the context that somewhere within an infinite span of existence, something apparently made God realize the need for independent beings capable of reciprocating his love.

Mercy and grace are attributes of God that he reveals because of his love toward his soul children and his desire for them to recognize that they have wandered far from him. Mercy is the expression of love that allowed God to take pity on his rebellious children. Grace is the expression of love that allowed God to establish a plan for the reconciliation of God and souls. The mind of each soul, created in the image of God, was originally in a state of mental attunement and harmony with the Mind of God. Soon after their creation, many souls began to exercise their free will to direct their minds toward desires and activities that were not in accordance with the plan or purpose of God. Souls were choosing selfishness above loving companionship with God. This condition would ultimately lead the soul mind to a state of separation from and loss of awareness of God, loss of the original conscious spiritual connection it had with God, and perhaps ultimately to the death of the soul.

God could have let these wayward souls remain in that condition without hope of returning to the original state of harmony and companionship with him. He didn't have to give errant souls the opportunity to recognize their error and change their ways. He could have focused his attention on the souls that remained true to him and let the others go their merry way into a self-imposed exile. Perhaps he could have set conditions and stipulations designed to chastise or punish rebellious souls for their misbehavior in the hope of affecting the desired change. Perhaps he could have altered the minds of the delinquent souls so that they would be unable to make decisions that went against his purpose for their creation. We will understand why this was not an option when we look at the nature of the soul. Instead, God provided a means for wayward souls to once again become aware of his presence, to feel the flow of his love, and regain consciousness of the fact that he is the Creator and Father.

> *The Lord is God of the universe, wherever thou art! For each soul finds self in that place which it occupies in the present only by the grace of God. [Edgar Cayce reading 3356-1]*

In Christian study guides one can find various statements defining grace as "the factor that opens our eyes to God's love," and as a "gift of awareness that enables us to see that we are out of alignment with God." One statement declares that "repentance is the outcome of grace." These particular statements were made in the context of a series of different kinds of grace that God bestows on us at different times. These and other definitions of grace broadly agree with grace as mentioned in the readings but don't have the same depth of meaning. Most of them seem to describe a positive beneficial action by God in response to a sin that the sinner has recognized and for which the sinner has repented. When a sinner recognizes his error and asks for forgiveness, God willingly grants the sinner an unmerited favor, specifically forgiveness of the sin. The readings recognize grace within a similar context, but they also see grace as encompassing the entirety of the purpose of the physical universe, the creation of human bodies, and the reason behind soul descent into materiality and their animation of human bodies. God is expressing grace by using materiality as an environment in which the soul can discover the damage its selfishness has caused to its relationship with God and reclaim its former glory with God. Grace extends across all human activity and all soul incarnations and is not something doled out on a case-by-case basis as a person asks forgiveness for a specific sin.

> *For, each soul, each entity, enters not merely by chance but through the grace, the mercy of a loving Father; that the soul may in and through its own choice work out of those faults, those fancies, as would prevent its communion wholly with - and an atonement with - the Creative Forces. [Edgar Cayce reading 459-12]*

Jesus explained the concept of divine love and grace in the parable of the prodigal son (Luke 15:11–32). In the story, a rich man's younger son asks for his share of the inheritance and takes leave of his family to strike out on his own. The son wastes his inherence on riotous and licentious living, eventually becoming destitute and nearly starving, reduced to tending and feeding swine, until he realizes that he is worse off than the lowliest servant in his father's house. Not until he reaches rock bottom and feels the maximum effect of his poor and selfish choices does the younger son fully understand his mistake in leaving the loving presence of his father and the comfort of his father's house. The suffering that the son endures while away from his father and the collapse of his life around him are the catalysts that make him aware of his mistakes. When the son finally finds the courage to admit his error and return to his family, he is greeted by a jubilant father and angry older son. The older son had been left to take care of the family business, doing all of the hard work that formerly was (presumably) shared by the sons. The parable makes it clear that the only thing of real importance to the father is that the errant younger son realizes that his separation from the family was the source of his downfall, will repent of his wrongdoing, and will willingly come back to the father. The son discovers that the freedom to express his independence in his own selfish way can have a negative effect; selfish expressions of will lead to separation from the father and family and descent into a living hell, while loving expressions of will lead to familial harmony and close companionship with the father.

Grace in this parable is the patient father waiting for the son to realize his mistake through activity in a harsh world of his own creation. When the son understands what he has lost and returns to the father, the father meets him with open arms. The father is free from anger and resentment at the younger son's actions and is purely joyful that the son has recognized that his downfall was caused by his own poor choices and that he decided to return. Mercy in this parable is the father's unwillingness to entertain any thought of punishment, although we may think the son richly deserved some kind of disciplinary action. There is no plan to put the son to work at the most demeaning and difficult field jobs until he has repaid the inheritance he has squandered. There are no slights and slurs spoken out of lingering bitterness and animosity toward the son. The father loves the son, and the son is back in the family and that is all that matters. There are other interesting elements in the parable. The son's selfish action in leaving his father and his quest for happiness through the pursuit of carnal desires and materialism in the far off land to which he traveled is his downfall. He eventually understands that he cannot find happiness by turning his back on his father and cannot find it in pursuit of material pleasure and self-gratification. The father does not chase after the younger son but is always there to support him after he decides to return. Of course, the parable is more than a story about a man and his relationship with his father. It is also an allegory about our soul, the rebellion in spirit of our soul against God, and our subsequent descent into the earth in quest of the illusive happiness we can never find here.

There is always the danger of reading too much into a parable but this parable well presents several facets of spiritual truth. The only aspect of the story that appears to conflict with its interpretation as an allegory for soul rebellion and redemption is the role played by the elder brother. The description in the readings of the loss of God awareness by souls who become too immersed in their own selfish activities is well represented by the younger son in the parable. Of course, the father in the parable represents God, the Father of each soul. But the readings identify the soul of Jesus as an elder brother to all other souls, state that his soul remained faithful to the Father, and credit the soul of Jesus with risking its spiritual status by incarnating into the earth to lead a rescue mission to the souls that had become trapped in materiality and had lost awareness of the way out of their self-imposed mess. In this sense, the angry elder brother of the parable is not a proper representation of God's family or the nature and activity of the soul of Jesus.

What is the purpose of grace and mercy on a personal level if we are destined to meet our spiritual shortcomings and failures through our physical relationships and material circumstances in a created physical universe? Sometimes our actions have placed us in such dire straits mentally and physically that even when we come to our spiritual senses and honestly desire to change our lives, the challenges before us seem too great to overcome. Discouragement over the severity and magnitude of the physical conditions we have created for ourselves and the mental effort necessary to overcome them can cause well-meant intentions to be crushed by the challenge and burden of our self-centered conscious desires, negative mental habits, and constant demands of our undisciplined mind for material goods and carnal pleasures. Out of his abundant love and mercy, God does not let us be tested beyond that which we are able to endure, which means we can face our many failures and their consequences in a manageable sequence of lives and life events. And of course, we always have available to us the added encouragement and strength that is found in the words and example of Jesus and the support of the Holy Spirit, if we are willing to sincerely ask for that help and are receptive to it.

God does not test us in the sense that traditional Christian writings and study guides often tell us. He is not looking for ways to trip us up to see how well we handle adversity. He does not let some evil spiritual being try to trick us into sinning. Every tribulation we face is created by our own previous actions and can be used as a positive experience that will bring spiritual growth to our soul as long as that life situation is approached by application of the fruits of the spirit with the faith and certainty that God will see us through to the end and will ultimately bring good out of a seemingly bad situation. God expresses his love for us by letting our actions reflect back upon us so we can see and eventually understand the error of our ways in the hope that we will turn back to him and accept the love he desires to pour upon us. It is not his will that any soul should perish (2 Peter 3:9), but with every failure of the soul to live up to its potential, he offers a way for it to recognize its failure and change its patterns of thought and activity. When we make a poor choice, a choice inconsistent with love and harmony, we create a ripple in the universe and our soul consciousness that will ultimately lead to yet another unpleasant future situation to confront. This cycle will only stop when we consistently make good, loving choices in our personal, financial, and business relationships.

Why are some people born into Christian families and exposed to Christian teaching and values while so many others are born into families who are Christian only in the sense that they obtained the label through parental association? Why are many born and raised with no, or limited, knowledge of Christianity and no opportunity to understand the mission and nature of Jesus and the hope that his message can bring? Why do so many Christians feel the need to consign the latter category of people to the trash bin of those going to hell because they "don't know Jesus?" All humans are his children, no matter what religion they profess or whether they were born into a stone-age society or technologically advanced society. Behind every human is a soul that God created and that he loves dearly. God does not favor any individual because he was born in a Christian society or was taken to church by his parents throughout his childhood. Being born in a Christian society or raised in a Christian denomination does not infer special rights or privileges upon any person. It does give each person who claims to be a Christian the attendant responsibilities to live up to that claim. This negative attitude toward adherents of other religions is not unique to Christians and is not a recent phenomenon. Peter was quite surprised when he realized that God is considerate of the spiritual welfare of all people, whether or not they were Jews.

> *And Cornelius said, … Now therefore are we all here present before God, to hear all things that are commanded thee of God. Then Peter opened his mouth, and said, Of a truth I perceive that God is no respecter of persons: But in every nation he that feareth him, and worketh righteousness, is accepted with him. (Acts 10:30-35)*

Make no mistake about it; this does not diminish the role of Jesus in the salvation of humanity. The readings state that the soul of Jesus has been actively involved, either directly as an incarnated individual or indirectly through the working of his Spirit on the earth, in every religious system that has taught God is One (364-9). All persons are treated equally before God, and all persons are given equal opportunity to seek God within the context of spiritual laws that operate impartially for all persons of every religion or creed. Every citizen of every country and every member of every religion has the opportunity to grow mentally and spiritually if they are willing to strive to make every thought and action conform to the commandments to love God and love neighbor. Love is the key factor in soul growth, not religion, doctrine, or creed. All souls enter earth for spiritual growth, and through time and reincarnation, all souls will have the opportunity to discover God and understand the spiritual necessity of applying the message of Jesus.

The Bible states that God is no respecter of persons (Romans 2:11–15), meaning that he is not impressed by or swayed by one's status or position in society. This also means he doesn't favor one established religion or doctrine over another. He does not even favor Christians over non-Christians, but he may very well expect a more perfect life from a Christian who has learned of the sacrifice of Jesus, knows the teachings of Jesus and his two commandments to mankind, and knows the manner in which Jesus applied those teachings during his life and ministry. But intellectual knowledge of Jesus and regular attendance at a Christian church doesn't confer one iota of special status or treatment from God. Souls who were and are born into non-Christian societies did so by choice and spiritual need, not because they are less deserving of God's favor. No particular segment of humankind, no matter what religion or race they represent, consciously possesses the full truth of the divine nature of God and the proper relationship between man and God. No single religion has yet been developed that includes a comprehensive theology that contains the full truth of our spiritual nature and potential to be companions of God.

God does not pour out his love on Christians to the exclusion of the rest of the world. A common statement that makes its way around Christian sermons and Sunday School class discussions is that only those who "name the name of Jesus" can be saved (Acts 4:12). This is too often interpreted to mean that every person who has not had the opportunity to hear about Jesus and has not accepted him as their savior because they were born before his time on the earth or because they lived or live in a country where the words of Jesus were never or are not now taught will go to hell or in some way are relegated to second-class citizenship in heaven. This is an all too common, but incorrect, interpretation of the biblical verse. Instead, God works in and through all people of all time and in all places to guide those to him who seek knowledge of him and who do his will in their lives, which is to love and worship him and love and serve others. This is possible without knowledge of the physical man Jesus because the Spirit of God has been active in all ages in directing and guiding the spiritual activity of humanity. God loves every soul he created and every human that is and has been animated by a soul.

The commandment of Jesus to love our neighbor as ourselves is not grounded in a feel-good philosophy of life or even in a sense of moral righteousness that God wants us to project toward others. Instead, it is based on the fact that every person on the earth is animated by a soul that was created by God and that every person on this earth is struggling to understand the true meaning of God and their proper relationship with God. Many of them have glimpsed the possibility of communion with God and are trying to express their current understanding of God as a loving Father through the way they live their lives, especially as it relates to interaction with others in the family, workplace, or community. Most of us fall far short of the moral standards we desire to express as Christians or as non-Christians, but it is easier to recognize the shortcomings of another than the shortcomings in one's own actions. There are some who have wandered so far from the presence of God that they no longer have a moral compass and are willing to prey on and misuse or abuse others to acquire material goods or to have the rush that can come from a feeling of power over others. It is much easier to hate or despise these people because their failures are so obvious, so public, and too often so despicable, but that would be equally wrong.

From the point of view of the readings, these persons are to be pitied rather than despised and are to be offered opportunities to change their mindset and reform their actions. This is part of the responsibility we have been given to be our brother's keeper. These people are guided by souls that were created in the same way and with the same potentiality and abilities of any other soul. But they have lost their way and turned their backs to the spiritual light that radiates from God and now live in a self-created darkness where it is difficult for them to find their way out. They are not unique, and they are not that different from the rest of us. In fact, at some point in the past or in a previous life we may have been just as lost and our actions may have been just as uncaring or irresponsible or criminal. It is difficult for many of us to imagine this possibility in the self-assurance of our certain salvation, the comfort of our material lives, or in our moral indignation over the behavior of others. But consider the experiences of Edgar Cayce himself. His life was as good and moral as is commonly seen among men. His sincere and selfless desire to be of service to God and his mental attunement to God permitted him to bring forth a treasure trove of medical and spiritual information that advanced the knowledge of the human race and placed religious thought on a more solid foundation. Yet, in a recent previous lifetime, he wasted much of his time on the earth seeking physical pleasure and enjoying a life of debauchery as John Bainbridge in early America. Although it is difficult to believe, those around us who behave poorly are just as much God's children and have just as much potential to do good as those who practice a virtuous lifestyle.

The teachings of Jesus are not to be understood to mean that we humans should congregate into groups or clubs of persons having similar ideas, who believe in similar partial truths, with special code words or secret handshakes to identify and separate ourselves from others. We are not to gloat over how much better we can interpret and understand his teachings than our fellow man. Jesus asked only two things of us, and all of his teachings were meant to expound upon and illustrate the spiritual value of those two important moral codes. His thoughts were in accordance with those teachings, his words explained those teachings, and his actions demonstrated how we should put those teachings into practice. The first teaching is to love God with all our heart, mind, soul, and strength. His second teaching is to love our neighbor as ourselves. God is love, and we are to respond to him in kind and with the fullness of our being, from and through our emotions, from the depths of our soul where our spiritual source is located and where we are closest to God, from our mind by desiring to willingly be directed and guided by him, and from our bodies, which are to be dedicated to serve others by manifesting the fruits of the spirit in all that we do. God as love expects us to generously express love.

Scientists have a difficult time studying love. It cannot be quantified or measured like a physical object. It can't be dissected like a biological specimen or extracted from the human body like drawing blood, yet it can have a tremendous impact on the soul individuality, human personality, and personal relationships. Some scientific theories have attempted to reduce love to genetics in the belief that all human traits can be traced back to DNA or to a Darwinian action active in early man to ensure the survival of the species. In the biological genetic interpretation of love and kindness, those human traits are hypothesized to originate deep in the ancestry of the human species (as hunters and gatherers) as a conscious desire to build reciprocity, so when we need help, others will come to our aid. Because we are likely to see others again during our foraging activities, we will be kind to them with the expectation of getting kindness in return later. This trait is now embedded in our genetic code. This hypothesis reduces love and kindness to a form of genetically generated manipulation of our fellow man. All actions of kindness and love arise out of selfishness, and unconditional, altruistic love does not exist. These actions are also postulated to induce a pleasurable feeling from dopamine release in the brain, which creates a desire to repeat the action. Neither scripture nor the readings support the idea that love, kindness, and other expressions of goodness toward our neighbor originate out of genetics or are inherited traits. The Bible and the secular history of man are full of stories where the sons of good men appear to have no concept of love or decency. The readings go so far as to declare that God created souls so that he could share his creative abilities and perhaps develop a real friendship and maybe even a sense of camaraderie with them. That may seem like blasphemy to many Christians who think we will spend our eternal existence in heaven in perpetual praise and worship of God, perhaps like a very long church service, which seems like the opposite extreme from the attempt by some scientists to explain love as genetics.

Why would an all-powerful creative being whose fundamental and intrinsic nature is love instill his most sociable divine quality into other spiritual beings and then not offer them the opportunity to fully express that quality? God did not seek to express the love that is an innate and inherent characteristic of his being in some abstract or metaphysical way unfathomable to lesser beings. Love needs something to be loved and to reciprocate that love, so that love might be satisfying and fulfilling. God desires thinking and feeling companions who are capable of responding to his generous love and with whom he can share his creative power and genius. Within the context of love, any true companion of God must be capable of interacting mentally and emotionally with him and, by extension, with each other. His companions must be able to comprehend the emotions that love evokes and be able to express and reciprocate that love. He is waiting for us to realize that human love, as special and fulfilling as it can be, is a poor substitute for the love that can exist between a soul and its Creator.

God is Law

> God as Law is fulfilled in the creation of the material universe and the personal and societal activity of humankind as directed by incarnated souls seeking opportunity to express their current understanding of God and shed the rebellious nature that led them to a state of separation from God.

For the LAW is set - and it happens! though a soul may will itself NEVER to reincarnate, but must burn and burn and burn - or suffer and suffer and suffer! For, the heaven and hell is built by the soul! The companionship in God is being one with Him; and the gift of God is being conscious of being one with Him, yet apart from Him - or one with, yet apart from, the Whole. [Edgar Cayce reading 5753-1]

God as law defines the rules and boundaries within which souls must operate. Whereas physical laws define the rules that govern the behavior of matter and energy and cannot be violated, God as Law is more like a legal system whereby actions outside of the law are met with immutable consequences. God's spiritual laws influence every activity and experience of a soul and set the arena within which it can move and interact with other souls. Spiritual laws do not violate our free will but do determine a future reward or future repercussion for each choice we make using our free will. The indwelling of God's love can influence the response of a soul to a life event, but God's laws do not influence the response of a soul to a life event. Souls choose to obey or disobey the law, but in the spiritual realm, they are often unaware of the detrimental effect of disobedience. The ultimate purpose of every soul active on the earth is to once again become aware of God as its Creator and to return to a loving relationship with him. Our return to a state of oneness and companionship with God, or salvation in Christian theology, does not depend entirely on him. We are responsible in large measure for our salvation, and we are entirely responsible for our spiritual isolation and life's trials and tribulations that too often buffet us.

What do we think when we hear God is Law? Do we think of Moses with long, flowing white hair standing on a mountain top holding a couple of stone tablets with the Ten Commandments inscribed on them? Do we imagine God as being clothed in a dark robe, sitting on a raised platform, looking down on the mass of newly dead, pronouncing judgment and dispensing justice? Is God's Law the 613 Jewish laws (*mitzvot* in Hebrew) that cover a multitude of personal dos and don'ts, ways to act or not act under various circumstances, and ways to levy justice upon those whose actions don't comply with the expected and required norm? In a legal sense, every sin is a crime perpetrated against God and is a thought or action that is in opposition to the spiritual rules of behavior that God put into effect for the society of souls and for the physical bodies they inhabit while expressing themselves on the earth. Man has considerable difficulty coming to agreement on exactly what does and does not constitute a sin. Although most people might agree that a murderer is a sinner, fewer these days would label a young dancing couple as sinners. What do the readings mean when they say God is Law? How can a spiritual being be a set of rules? What do the readings mean by the statement that Jesus became the Law? Is this just an alternate way of expressing the perfection of Jesus and the oneness between God and Jesus, or is there a deeper meaning? We know what man's legal systems look like, but what does God's legal system look like?

Legal systems set the rules of behavior for man in society and describe the punishments to be administered when those rules are broken. Justice is the fair and impartial administration of the law. The concept of justice means a right punishment for a particular crime, usually based on the perceived severity of the crime, the amount of physical damage done to property, or the amount of physical, mental, and emotional damage inflicted on another individual. The legal system and the manner in which justice is applied varies from country to country according to legal tradition and religious influence in the government, but the idea that a criminal needs to be punished to discourage future aggressive actions by the perpetrator and to deter and prevent other members of society from engaging in the same behavior is common to all of them.

Christian scriptures record instances of brutal forms of punishment for what today seem like minor offenses. The use of whips as a form of attitude adjustment and slow, painful suffocation on a cross, perhaps facilitated by a pair of broken femurs, as a means of inflicting capital punishment have mostly gone by the wayside, although there is the occasional despot in certain authoritarian societies that tries to surreptitiously reinstate similar methods of control and punishment. Some Christian nations still inflict harsh punishment for certain capital crimes but usually try to minimize the inflicted pain when carrying out death sentences or long periods of imprisonment. Western societies have largely transitioned away from harsh punishment for behavior such as adultery and homosexuality, treating them more as civil matters and personal choices instead of crimes, but some countries still stone to death women caught in adulterous situations, amputate hands as a deterrent to thieves, and harass, discriminate against, and kill homosexuals. The harsher punishments are often, but not always, meted out in countries where there is no real concept of separation of church and state, and religion plays a major role in defining the government's legal system, such as in the Islamic countries of the Middle East. One could wonder why intense religious piety seems to go hand-in-hand with more brutal and harsher criminal punishments.

Jesus didn't teach or endorse harsh measures such as death by stoning as a punishment for adultery—even though the Old Testament, essentially a Jewish scripture, calls for adulterers to be put to death (Leviticus 20:10–12)—nor did he advocate for the repeal of such laws. He didn't discuss homosexuality at all, but many Christians fall back on Jewish Old Testament passages (Leviticus 18:22 and Leviticus 20:13) and the words of Paul (Romans 1:26–27) to build their understanding of the legal and moral status of homosexuals. Why do many Christians today actively and strongly turn to the Bible to find support for harsh measures against homosexuals (Hamilton, 2013) and ignore its passages that demand death for adulterers? Jesus took a different approach. He tended to focus on forgiveness and encouraging offenders to change their ways (John 7:53–8:11). He realized that man will be spiritually transformed only when he seeks to love God and neighbor whole heartedly, without reservation and with an attitude that lets God be the driving force within his or her life instead of the self-centered attitude held by much of humanity.

> *Each nation has set some standard of some activity of man as its idea, either of man's keeping himself for himself or of those in such other nations as man's preparation for that companionship with God. For remember, there are unchangeable laws. For God is law. Law is God. Love is law. Love is God. There are then in the hearts, the minds of man, various concepts of these laws and as to where and to what they are applicable. [Edgar Cayce reading 3976-29]*

Most Christians believe that God's mercy is available to any repentant sinner who accepts Christ's (a misuse of the term) atoning death and asks for forgiveness (1 Corinthians 15:3–4). Unrepentant sinners get punished by being consigned to hell and God's eternal wrath after death because of their unwillingness to repent and change their behavior. Some Christians even believe that none of it really matters because God has already preordained that a certain fraction of the human population will be granted the right to enter heaven, and he has already completed the selection process. God, as absolute supreme ruler, ultimately decides who will receive his mercy and who will receive his wrath. Of course, most of these people believe that they are part of the chosen few. Consider a hypothetical life situation. A young woman is brutally beaten and raped by a man whose identity is never discovered. Over time, as he grows older, the man feels true remorse, repents and asks God for forgiveness, and settles into society as a model citizen. God knows the man's heart has changed and in his mercy accepts him into heaven when he dies. Meanwhile, the woman is so traumatized by the attack that she loses all sense of self-respect and descends into a deep and lasting depression. She strikes out in anger at friends and family, stops attending social events and the church she was thinking about joining, resorts to drugs to handle the pain and shame, and eventually dies of a drug overdose, alone and all but forgotten. As an unrepentant sinner and drug addict, God sends her where she deserves to be: to hell, where her torment is magnified and she can spend eternity agonizing over her failure to put her life back in order. It is difficult to imagine how mercy and justice are properly dispensed in this hypothetical, but not impossible or improbable, scenario.

Christians want desperately to believe that confession and absolution are two sides of an infinitely thin coin. The act of confession immediately grants the confessor a full pardon, absolution from the sin, and clemency from any responsibility for the damage the sin caused to self and others. The concept of restitution is foreign to Christian reconciliation and salvation. The forgiveness of sin, whether received as part of the Sacrament of Penance or by individual prayer and repentance as part of a personal and heart-felt appeal to God for mercy involves the penitent, not the victim. A priest spends a lifetime enjoying the company of young boys in his parish and tops it off by confessing and receiving absolution in his declining days, the icing on the cake. The physical and emotional damage caused to the victims is not addressed by religious authorities, and in most cases the priest's activities are kept hidden for fear of damaging or soiling the good name of the religious institution unless the authorities are compelled to reveal them by secular authorities and public opinion. A Mafia don orders a series of murders in his heyday to consolidate his criminal empire but has a change of heart in his later years. No problem. A sincere confession and a display of true remorse, and a new man is formed, as innocent and pure as the day he was born, and the path to heaven is swept free of obstacles. No one mentions or remembers the victims of his criminal activities that died because of his direct action or by his order to his henchmen. Despite the assertion of priests and pastors and the confidence of Christians, deathbed conversions do not erase a lifetime of sin and grant access to heaven to the sinner. The act of repentance can begin the process whereby a soul embarks on a path that eventually leads it to closer companionship with God, but it is not the end of that path.

Many Christians emphatically believe that salvation is by faith through the grace of God and that the works of individuals, although good as social exercise and as an outward expression of a loving Christian attitude, cannot bring anyone to salvation. The apostles Paul and James discussed these issues, with Paul putting more emphasis on justification before God by faith and grace (Galatians 2:16) and James emphasizing that works are a natural and necessary extension of faith (James 1:22–25). Grace is too often interpreted in Christianity as a grant of immunity from our actions given to us by God when we recognize our actions are wrong and repent of them. This is a false concept. All thoughts and actions have consequences. Grace does not interrupt or negate the fulfillment of spiritual law. The maxim that we reap what we sow is an expression of the most basic spiritual law. Actions, manifested thoughts, are causal in the physical world and have real-world and spiritual consequences.

The readings state that each soul occupies its place in a world filled with opportunities for the soul to make choices that can lead to fellowship and companionship with God, and they emphasize that faith opens the mind to the true meaning of life and allows the soul to act in accordance with a higher spiritual purpose. The reason for these life opportunities is to allow the soul to express its understanding of God through its natural spiritual attributes, an activity that draws it closer to God. Grace isn't the action by which God forgives a sin or grants salvation. Grace means that God continues to offer us daily opportunities to change even though we are undeserving of those opportunities. The physical universe and soul incarnation in human bodies are manifested grace. Good works are the material expression of an abiding love of God within the soul and are manifested in the world through the application of the fruits of the spirit (kindness, love, patience, long suffering, etc.) in interpersonal relationships. Good works are an outward expression of an inner alignment of the soul with its creator. Good works help reawaken the soul mind to the sure knowledge of its compassionate and loving Father. The soul works on meeting its shortcomings before God by having a better, more loving attitude when dealing with other incarnated souls and through service to fellow incarnated souls. We are judged by the works we do (Revelations 20:12 and Hebrews 9:27) because material activity arises from a mind that seeks to please God or to glorify self. It is counted to our credit when our works are the actions that naturally flow from selflessly loving and serving our neighbor.

So, what do most Christians imagine God's legal system and form of justice looks like? It looks like the Christian teaching of salvation. Perhaps we don't think of salvation as God's legal system, but it seems to have all the essential characteristics of a legal system. There is a set of rules and a defined reward for adhering to them and an established punishment for breaking them. The doctrine of salvation may vary a little from one Christian denomination to another but, to summarize, it basically goes like this. We are naturally sinful human beings or prone to sin because of Adam's fall from grace and as such are not worthy of a place in heaven with God. Our sins are, in essence, crimes against God. If we obey the rules by remaining sinless, we can expect to be rewarded with a place in heaven. When we chose to commit sins and remain in a state of sin, as demonstrated by repeated actions in defiance of the commandments of Jesus, we must incur a penalty for our misbehavior. When we continue to break the rules without expressing sorrow for our actions before we die, the penalty is some form of eternal torture by fire in the depths of hell or in its milder form, eternal isolation and banishment from God and from the realm of heaven. Left to ourselves, we are totally incapable of extracting ourselves from this sinful condition or of earning our salvation through any action or activity we undertake. It is only by God's grace that we are saved from our sins. Good deeds (works) in themselves do not gain us any merit with God or sway his decision concerning our salvation, although there is some question and controversy about the value of works if they are a natural extension of faith in God.

Because we are God's children and he loves us, he has prepared a way for us to be saved from our sins. To help us out of this self-inflicted mess, God sent his son Jesus to earth (or in some versions, God came to earth in human form as Jesus) to take on our sins and suffer and die for us on a cross. There are variations on the theme, but in general the method of escape from our debased condition is to recognize that we are sinners and realize that only God has the ability to save us from our sins (faith), to confess our sins and change the lifestyle we are living by renouncing and turning away from our sinful actions (repentance), to accept Jesus as our personal savior, and to ask God to pardon us for our previous sinful behavior (forgiveness). When we sincerely do this, God in his infinite grace grants us absolution from all previous sins and guarantees that he will receive us after death into a special place he has prepared for us in heaven. Every sinner can apply for and receive this absolution, without incurring punishment for their prior actions, at any time of their choosing during their lifetime. These perpetrators, criminals in the legal system of salvation, are rewarded for coming forward and repenting without regard to the pain and suffering their sinful actions brought to others. The action of repentance and acceptance of salvation not only absolves the sinner from penalty, it also elevates the spiritual social status of the individual and allows him access to a piece of prime spiritual real estate with an eternal lease.

There are no caveats and no extenuating circumstances. Once the change of heart is confirmed and verified by God, it is a done deal. Jesus (the lawyer) works out a plea bargain for his client (the soul) with God (the judge) to reduce the sentence from eternal punishment to time spent on the earth in a physical body. The heavenly reward may be delayed momentarily so that everyone who is born between now and some indefinite future time also gets a chance to accept the free offer of salvation, but in the end all who do accept it will be raised from the dead and given a new and better body that will last forever. As long as we keep our part of the bargain, God will keep his. God, in his infinite grace, is ready and willing to grant absolution and full pardon to any sinner for any sin committed at any time during his lifetime. All we need to do on our part is to ask for his help, but if we procrastinate to the point of death, it is too late, and the offer expires.

Most of us procrastinate at some time or another over something we need to do but don't want to do. We keep finding excuses to put it off. Many times our excuses coalesce around the idea that we haven't yet found the time to get it done, it will require more effort than we are willing to expend at the moment, or it will interrupt our present schedule of activities so should be put off until a later, more convenient date. But procrastination really is not too much of a problem when it comes to salvation as long as the process of repentance and forgiveness is completed before death. A deathbed confession and repentance along with a statement of belief in Jesus can obliterate the consequences of a lifetime of sin. Also, under God's system of justice that we call salvation, there are no worries about restoration payments to the victims, even victims of our more serious and vicious criminal sins. In fact, the victim isn't even considered in God's legal system of salvation. There is no concept of justice for the victim. Traditional Christian churches accept and embrace this doctrine, declaring it to be the concept of justice established by God to save human beings from the consequences of their sins. Surely, there is no better system of justice than the one created by and used by God. The real question is this: if Christians actually believe that this doctrine is the form of justice God uses and prefers and has established for the salvation of humankind, then why don't Christian nations model their legal systems on this more perfect system of justice?

Think of the savings society would incur if a system of justice based on the Christian concept of salvation were fully adopted. The budget and personnel of police departments and courts could be slashed to the bare bones, which would substantially reduce taxes. Crimes would not be investigated, and the police would not interfere with criminal activities because criminals would be given the freedom to do as they pleased until they decided that their behavior was wrong and made the personal decision to change. Repentant criminals could come to a bare-bones court system to confess their crimes, but no jails would be needed to hold criminals before or after trial. Criminals who realize and accept that their crimes against the community are morally wrong and violate the accepted norms of behavior and the laws of society, who confess their crimes before the judge in open court, and who repent of their criminal conduct are automatically granted immunity from punishment. They are unconditionally accepted back into the community they preyed upon with rejoicing and open arms. As a bonus, they will qualify for housing in the most prestigious part of town. Petty theft or murder; there is no consideration of punishment or restitution. Backsliding usually poses no problem as long as the criminal appears before the court after their latest crime spree and again sincerely asks for forgiveness. After an allotted period of time unknown to them, criminals who choose not to confess their crimes before the court or who repent repeatedly, especially for serious crimes, but fail to change their behavior are written off as hopeless cases. They are sent to a special facility where they undergo an especially gruesome and painful form of torture just short of death that includes such favorites as slow roasting over a large fire or, if they are more fortunate, are banished to a remote island with the other unrepentant criminals and are denied contact with the rest of the world.

Justice in man's legal system usually focuses on punishment and to a lesser degree on rehabilitation. In man's legal system criminals are actively pursued and kept in check, but in a system patterned after Christian salvation, criminals would have the freedom to terrorize the community until they had a personal change of heart, until they heeded the pleas of the community to live within the law. The amount of suffering and property damage among victims likely would be far greater under a salvation system of justice because there would be no attempt to catch criminals in the act of committing a crime and no attempt to stop them from committing future crimes. As in the Christian understanding of sin and salvation, sinners against humanity—the perpetrators of criminal actions against society—would be left to ravage the society for as long as they desired within the boundary set by their life span.

If Christian salvation is the perfect dispensation of justice for the soul, then why don't we set up our societal justice systems to operate in the same way? Maybe because deep down inside we know that such a system doesn't make much sense and would be a disaster to any society that attempted to imitate it. Maybe our understanding of God's system of justice and salvation is flawed. Certainly God is much more focused on the immortal souls that animate men rather than the mortal bodies that are animated by souls. Maybe God actually has another system of justice in place that not only saves souls but makes souls accountable for their thoughts and actions and rectifies the physical, emotional, and mental damage caused by incarnated souls to their neighbors. Maybe the boundary between guilt and innocence and the dispensing of justice to souls is more complex than we imagine. The readings indicate that nothing major happens to us in this world without a reason, with the caveat that accidents can happen in materiality (364-6). The primary reason for immoral and criminal behavior and the suffering it causes is our separation from God and the poor and selfish choices we make as we try to go it alone. Maybe we need to start thinking of the world of causality in which we humans live as God's system of justice, a physical universe with the spiritual purpose of soul salvation.

The justice system that Christianity associates with salvation for repentant sinners and punishment for unrepentant sinners is attractive because it can provide a shortcut to heaven with minimal effort and immunity from the consequences of even our worst actions. We only need to be willing to take that first step on faith and reach out to him. We only need to face life once. But the seductive justice system of salvation leads us to ignore the scriptures that explain that we must reap what we sow (Galatians 6:7–8), that place the responsibility of our actions squarely on us as souls and that tell us we receive according to the manner in which we have given. Maybe deep down in our subconscious minds we do understand the pitfalls of salvation as currently taught. The concept of Christian salvation doesn't help us to understand the deeper underlying source and cause of human suffering. We can be quick to point to the suffering of others and explain it as being the result of sin in their lives, but we aren't as likely to point to ourselves and claim we are suffering because we are sinful. We never seem to be able come up with a reasonable explanation for why so many apparently "good" people suffer just as much and just as severely as so many "bad" people. A little humility and heartfelt repentance can absolve us from any and all retribution for anything from an angry word directed to a spouse to murder as long as we are willing to bring our case before God before we die, even if it is just a minute or two before we die. It is almost too good to be true.

> *Then know that love is law, and that law is that which may bring about the most necessary things in the mental, physical, and spiritual life of a body; for God will not be mocked by man's nor woman's, own insignificant ideas of self's importance as to laws concerning the mental, or the physical, or the spiritual being. [Edgar Cayce reading 349-6]*

The statement that salvation is by grace alone is sometimes used by Christians to declare that we can do nothing of our own accord to gain salvation; it is entirely in the hands of God, and we are powerless to do anything to facilitate our own salvation beyond faith, repentance, and living the best life we can while hoping for the best outcome. However, it is not a matter of earning salvation verses salvation being granted by the grace of God, it is a matter of souls being immersed in an environment that allows them to perceive the loss of their spiritual heritage and acquire knowledge and understanding of the new mental attitude needed to implement changes in consciousness and patterns of behavior. The physical, mental, and emotional environment that constitutes our life was created by us as the manifestation of previous activities by our soul in relation to spiritual law. The soul of every individual is required to make a decision and conscious effort to repeatedly and consistently apply spiritual principles during its incarnation that will lead it to a restored relationship with its creator.

Salvation is not a single event that takes place when one realizes and accepts that Jesus is the Son of God. Salvation is a process that begins when we recognize that Jesus is the Way and that progresses as we continue to live our life in accord with that Way. Salvation begins when we exercise faith and allow the physically invisible but mentally and spiritually available inner presence of God to work in our lives, when we obey the inner prompting that comes from that awareness of God's presence, and when we alter our outward behavior in response. Salvation can take place outside of the context of a physical Jesus and therefore has always been and is still available to everyone, Christian or non-Christian, who is willing to acknowledge the presence of God, open their heart and mind to accept guidance from that presence, and who is willing to turn away from sin and change their behavior. Jesus prepared the way by completing the process of mental and physical perfection on the earth, but each of us is required to walk that same path to our own perfection and so restore our broken relationship with God. Salvation begins when we become mentally attuned to the inner prompting from God that will direct our lives so that we yield the fruits of the spirit as a natural part of our daily activities. To believe in Jesus in the spiritual sense that leads to soul growth is to believe in his message and act accordingly, and his message was to follow him, follow his example by applying his two commandments in our lives.

> *For God so loved the world, that he gave his only begotten Son, that whosoever believeth in him should not perish, but have everlasting life. (John 3:16)*

> *… for be not deceived, God is not mocked, and what a body - mentally, physically, spiritually - sows, that it reaps! and not something else! [Edgar Cayce reading 349-4]*

So if God's legal system for dispensing justice isn't salvation as understood by traditional Christianity, then what do the readings mean by the words "God is Law"? To say Law and God are synonymous is not the same as saying God formulated a legal system to punish or reward sinners based on their willingness to report their crime and repent of their prior actions. One of the implications of God as Law is that God is not separate from the Law. In some way he is law itself, both in promise and in action. The scriptures and the readings state that God is no respecter of persons (Romans 2:11–15), and this aspect of God's nature allows him to apply the law fairly and impartially for every soul, every human being, no matter how high or low they are on the social ladder of their society. Just like physical law and man's legal system, spiritual law operates whether or not we understand it, acknowledge it, or ignore it. God does not use the Christian system of salvation to pass judgment and enforce justice. God's real legal system is an impersonal, impartial, and fair system that is incorporated into the physical structure and causal nature of the universe and is designed to redeem and save souls, not punish them. Hebrews 9:27 states that, "it is appointed unto men once to die, but after this the judgment." It can be taken as support for the traditional idea of salvation, but that would be incorrect. The judgment of the soul at the end of each individual human life (not at the end of only one life for each human) is to assess the progress of the soul by reviewing what the soul did with the opportunities it was offered during its experience on the earth and to determine the next step that will best prepare the soul for companionship with God.

God as Law is the driving force behind the creation of the physical universe, the earth that revolves around an obscure sun at the edge of a galaxy of no special importance, and all of the diverse biological life that has ever walked across its surface, flown through its atmosphere, or swam through its waters. This is the environment that sets souls free and that offers them the opportunity to become reacquainted and reunited with their estranged Father. The universe does not exist for the pleasure of man or as an arena where God can observe our activities and tally or register our successes and failures at keeping his laws. It is Law itself. It exists for the benefit of the souls that occupy the bodies of men and women, souls that have spent far too long away from their Father and have lost the ability to communicate with him and have even lost knowledge of the true system of salvation. How can that be? How can a physical universe be a reeducation center and rehabilitation facility for souls, spiritual beings whose natural environment probably has little in common with the world we experience through physical bodies? Each time we encounter another incarnated soul on a deep emotional level during our earthly lives, the place and circumstance of that encounter and our spiritual consciousness or awareness at that moment is precisely what is needed to spiritually advance the soul. However, spiritual advancement will only occur if the soul is willing to learn to repeatedly make choices during its experiences that are consistent with what it sincerely believes God would wish it to do within the context of loving God and loving thy neighbor as thyself.

God is Life

God as Life is seen as the organizing principle that aligns atoms in a crystal, brings organic molecules together to form DNA, RNA, amino acids and proteins and choreographs the parts into a unified whole from the tiniest cell to the most complex higher organism.

See HIM [God] as He manifests in every form of life; for He IS Life in ALL its manifestations in the earth! [Edgar Cayce reading 488-6]

... whether it is mineral, vegetable or animal, these are spiritualized in that ability of using, doing, being all that the Creator had given them to do. [Edgar Cayce reading 281-54]

All are of one, as all are in the Mind of the Creator, all made through the Mind of the Creator, to serve in their various capacities of executing the homage, service, to the Creator. [Edgar Cayce reading 900-89]

God is the creator of nonphysical life in the spiritual realm, including the host of angelic beings with their hierarchy and separate purposes, and the creator of souls as companions with the capacity to love and ability to share in the joy of his creative force. God gave life to souls because he needed to express his boundless love. He is the author of all physical life, from the most primitive cell to the human body. Because God is One, all forms of life—whether pure spirit or spirit-directed biological life—are a portion of God, are animated by the Mind of God, and are active participants in his creative desires. God created the human form so that incarnated souls could learn to express unselfish love through the activity of the human body in its relationship with other incarnated souls. Souls can also express love toward creatures of the plant and animal kingdom, but only higher-order animals can respond to and share that love at a basic level. Spirit and mind are necessary to animate biological life. Will, the aspect of a soul that gives it independence and the freedom to make higher-level choices, makes the human life form different from other animal life forms. Animals do not have souls; their bodies are only animated by mind and spirit. They do not have independent will or the associated independent mental faculty of the soul known as the subconscious mind (900-31).

Because animals do not have the free will that is necessary to allow them to choose or reject companionship with God, they lack the ability to defy their creator. Although they are animated by the activity of mind in relation to spirit, they do not have the independent will that would allow them to direct mind into paths of expression that are inconsistent with the environmental niche that evolution designed them to fill. Their mental capabilities are mostly confined to the propagation of their species, the acquisition of food and shelter, and the application of survival instincts that assure the continuity of their species (262-63). Man has been given the ability to subdue the earth (Genesis 1:28), and that includes the ability to influence the mind of animals to encourage them to conform to his wishes. Unfortunately, man has yet to acquire and master the ability to control his own rebellious and obstinate mind. The human body, as a biological and animal life form, shares these instinctive animal traits, and the human mind can magnify them in consciousness until they become a major driving force within the body and draw the soul more deeply into the carnal environment, causing damage to the soul of the individual. The readings indicate that the ability of humans to fully show compassion for and care for animals and other earthly creatures will come naturally as man learns to love and care for other humans. Our callous disregard for plant and animal life is apparent in the rate at which we are sending species to extinction by destroying and disrupting their natural environments with our massive construction projects and by the unnatural epoch of global warming we are inflicting upon the earth, largely a byproduct of our incessant quest for material goods and wealth. Our sewage and trash chokes the life out of the great oceans while we drown in our ever-increasing use of plastic while whining that recycling is inconvenient. We want God to care for us, but in our turn we are unwilling to care for his other creations.

God as life is an extension or continuation of God as law. Law came from God in the creation of the fabric of spacetime, which is designed such that matter and energy interact in a causal relation where the manifested thoughts and activity of every incarnated soul precipitates a definite observable and measurable effect on other incarnated souls and material objects. God as law led to the creation of the universe as a suitable causal environment for souls to reconnect to their spiritual roots and have the opportunity to regain their spiritual heritage and the conscious knowledge of their favored status as children of God. The structure of spacetime, which encodes the physical laws of the universe, operates in conjunction with the spiritual laws that apply to souls to enforce the biblical axiom that every soul reaps what it sows while manifesting activity through the body (262-52, 257-170, and many others). But the creation of the physical universe through an evolutionary process involving energy-matter interactions and gravitational attraction of matter only created the earth as an inanimate ball of rock and water with an atmosphere that was toxic to most life as we know it today. It was not an environment conducive to the spiritual growth of souls.

The new universe provided a causal environment that applied to physical objects, those made of atomic matter subject to physical laws. Discarnate souls could recognize the planet, visit the planet, and perhaps even "walk" on the surface of the planet (or whatever passes for walking to a nonphysical being) if they wished, but there was nothing to attract them and hold their attention, no reason for them to take an interest in one particular planet among many others. Physical processes directed by universal laws had prepared the stage for the arrival of the main actors but had not yet created the supporting cast. God, as an impelling force of spirit and mind, activated primitive organic molecules and amino acids to set into motion the first primitive forms of the rich and diverse biological life that would be essential to complete his plan for a soul rehabilitation center.

Scientists have seen little direct evidence of the initial event that set the physical universe into motion and the first moment when a spark of life appeared on the earth, but they have produced several theories and have collected a considerable body of knowledge and observations relating to the beginning of the physical universe and the beginning of biological life on the earth. The closer their investigations get to these initial events, the more difficult it is to analyze, measure, and extrapolate current knowledge to the correct answers. Scientists do know that primitive life on the earth began about 3.8 Bya (billion years ago) and that it took billions of years for the biological environment to develop and mature to the point where it could sustain the higher-order biological life form we call human. Scientists do not recognize that this life form is able to accommodate soul activity so that it can be used by the soul to experience responsibility for its thoughts and actions and reawaken God consciousness.

For the purpose of scientific study, all forms of life can be thought of as organized organic molecules functioning as a cooperative unit; biological organisms that develop and change over time in response to environmental stresses. The cellular structure of each life form can be investigated, their organs can be identified, and their purpose and interrelation can be assessed, their body types can be categorized and arranged in family trees that indicate their genetic relationships and ancestry, and their patterns of behavior and social interactions can be observed and recorded. Their neural functions and brain activity can be subjected to experimental testing designed to deduce how they respond to stress to give insight into how the organism intuitively reacts to stimuli and perhaps how it thinks. Biological science has made great strides in recent decades, especially since the discovery of DNA and its role in cell reproduction and the passing of genetic material to offspring, but science still hasn't found irrefutable evidence of the source of biological life.

So, what is the source of the vast diversity of life we see on the earth? Each cell in every biological organism has a driving force, an animating life force that makes organic matter living matter. Plant and animal life in nature are driven by the instinct to acquire energy from their environment, to reproduce, and to preserve the integrity of their bodily form in the face of danger from competitors or their environment (262-63). The readings even go so far as to indicate that inanimate matter, in the form of minerals, are a rudimentary form of life, perhaps a transitional form between unanimated matter and animated matter.

The basic life force, the one common to mineral, plant, and animal, including humans, seems to be an organizing principle that guides collections of molecules to be something greater than their individual activity can achieve, much as God desires all souls to recognize their unity in him and become a greater unified whole than anything they can imagine as solitary units of spirit. It is organization and cooperation on a scale and complexity that is unimaginable to the finite human consciousness.

> *Then Life, or the manifestation of that which is in motion, is receiving its impulse from a first cause. [Edgar Cayce reading 254-67]*

From a physical and mechanistic viewpoint, science has determined that cellular biological activity is essentially the movement of ions (charged atoms or molecules) across the outer membrane of the cell and within the cell body. Neurons in the human body pass information as electrical impulses from sensory cells near the surface of the body to the brain and from the brain to muscles. The readings indicate that God, electricity, and life are intimately related in that God expresses himself in the physical universe through the activity of electrical force, the "pull and push" of electrically charged subatomic particles and charged atomic structures. The movement of ions in a cell is guided by electric potentials (charged particles move from regions of higher potential to regions of lower potential), and the distribution of ions determines the strength of the electric potential across a cell membrane. Since the readings intimately connect electrical energy with God, one might argue that science has already observed God in action as biological activity and just hasn't made the intellectual leap needed to recognize that connection. But this doesn't seem to be a complete explanation of God's role in biological life. Inorganic compounds also have electrical activity, but they don't organize to form anything we might consider to be cellular life. Also, electrical activity may sustain life processes but would seem to fall short when it comes to generating self-consciousness. The readings indicate that every cell of the human body has a rudimentary consciousness that must be raised to some threshold level as part of the process of spiritually raising the conscious and subconscious mind of the soul. If the readings are correct, scientists still have a lot of work to do to fully comprehend the nature of the body-mind connection.

What is the force that animates human life? The first thing to jump into the mind of a Christian might be passages in Genesis that describe God breathing "the breath of life" into man and man becoming "a living soul" (Genesis 2:7), but this does not fully answer the question. The first primitive human life forms preceded the current human life form by a few million years. Humans did not come into existence 6,000 years ago, and souls did not come into existence with the creation of Adam and Eve or at the moment of the birth of each of their many descendants. The readings indicate that modern human bodies (specifically, Adamic bodies created for the purpose of facilitating soul growth) were indeed prepared by God for occupation by souls. The readings are also clear that the biblical Adam and Eve story describes, in symbolic terms, the activity of two of these created beings but indicates that they are also archetypes of a large number of similar bodies used by incarnating souls. For a long time before the Adamic bodies came to dominate the human populations, souls experimented with materiality and forms of sensual self-gratification by intercepting sensory information from plant and nonhuman animal life forms. Souls were never the driving force behind the biological life that has thrived on the earth for billions of years, although they did interfere with the evolution of some of these life forms. Those forms of experimentation and interference with plant and animal life were brought to an end with the arrival of the Adamic bodies, as indicated in the biblical Book of Genesis. Since that era, souls have only been able to interfere with the natural evolutionary processes of biological life indirectly through mind-directed actions of a human body (364-7). The human body today can be thought of as having two forms of life. The first, in common with other biological life forms, is the Spirit and Mind of God that impels the basic biological life force that directs and activates each cell of the body to fulfill its tiny part within the complexity of the whole organism. The second, unique to the human life form, is a soul mind that directs and controls the overall life activity of the body at the urging of the soul.

> *Life is creative, and is the manifestation of that energy, that oneness, which may never be WHOLLY discerned or discovered in materiality, - and yet is the basis of all motivative forces and influences in the experiences of an individual. [Edgar Cayce reading 2012-1]*

God is Truth

God as Truth is perfectly revealed in Christ Consciousness, the ideal mental state whereby our soul communicates directly and personally with God free from any distortions caused by egotism, material self-obsession or the influence of other individuals or organizations that claim to have inside knowledge to the mind and will of God.

What is Truth? That which makes aware to the inmost self or the soul the Divine and its purposes with that soul. [Edgar Cayce reading 262-81]

Then, as we see, the lessons as are gained, as the truths in science, as the truths in religion, to meet the needs of the thinking man must be compatible to each other, ... [Edgar Cayce reading 900-171]

What is truth? What defines or encapsulates the truth? We do not usually come to our concept of truth by inspiration or revelation but as a result of accumulated experiences and knowledge, the books that we have read that resonate with something deep inside us, the conversations we have with people we respect and admire, or perhaps those with whom we have little in common or epitomize the opposite of truth to us. An interesting phenomenon occurs among many Christians who profess to believe that the Bible contains the infallible and perpetual word of God and should be read as literal fact, as absolute truth; they read it and come to a different concept of truth than other Christians who read the same book. This is revealed in the different creeds and doctrines that have been promoted by different denominations over the centuries. However, too many Christians still think they know the truth to the extent and depth of conviction that it becomes their mission to impose their concept of truth upon their fellow man.

To deny the validity of long-established and verified scientific knowledge because it does not conform to a specific interpretation of the Bible promoted by a religious organization or respected religious figure does not contribute to the advancement of truth. The acceptance of a particular doctrine by a religion or church denomination does not make it truth but only a statement of faith or unsupported belief. Faith is an attribute of the soul that gives it the strength to see beyond the physical and perceive the inner presence of the Creator. Faith does not give any person special insight into the physical world of matter and energy or the nature of biological life. Much religious thought and interpretation has been wrong in the past and undoubtedly is still wrong in the present and unfortunately will continue to be wrong in the future. Adherents to some religious interpretations of Christian scripture are often reluctant to alter their theological views in the light of new ideas and revelations that arise from the scientific community. This misuse of faith can lead to self-inflicted ignorance. The desire to believe that something is true does not make it true, especially when that truth can be proven wrong by observation and unbiased investigation.

It is certain that a scientific theory can be wrong, especially at its inception before the theory and its predictions can be fully confirmed by experiment and peer review. But the use of the scientific method during the past 500 years has allowed the previous inconclusive and unproven ideas of natural philosophy to be replaced by rigorous mathematical theories that have been tested and verified (or rejected) by prediction and experimental measurements, most notably in the hard sciences of physics, chemistry, geology, and biology. This process of iterating a hypothesis or theory to a solid principle or physical law weeds out weaker and incorrect theories and gives a better understanding of the limitations of the stronger theories. Religious leaders that rail against the theory of natural selection and evolution, for example, are wasting a lot of spiritual capital. These theories will be studied, tested, modified, verified, or rejected and will iterate to a more accurate and correct version as time goes on with or without their interference or vocal declarations that the Bible proves the science wrong.

Biblical interpretation is not infallible, and the Bible was written by men who were a product of their culture and times, education or lack thereof, and whose spiritual attunement was not perfect. Most of the books of the Bible were written and compiled decades or centuries after the events they describe, and one of them describes events that occurred billions of years ago. God used the tools that were available to him, people willing to be used to bring the message of his existence and love for humankind to the world, but that does not mean that everyone who wrote the letters and manuscripts that were eventually compiled into the Bible had access to reliable sources of information, or individuals with perfect memories, or worked from perfect manuscripts or perfectly translated documents that correctly recorded the intended meanings of the authors. The teachings of the Bible and other scriptures should undergo a similar process of research and study as is given to scientific theories. This does not mean that belief in God or spiritual laws should be rejected because they have not been proven by science or are not fully compatible with scientific explanations of physical phenomena. There are many religious topics and ideas that do not intersect with science and do not lend themselves to investigation through the study of physical matter and energy, but the many pseudo-scientific claims about the physical universe derived from the literal interpretation of the Old Testament should be discarded because they do contradict known and verified scientific truths.

Spiritual laws primarily concern spiritual matters of the soul but also have a physical effect. This spiritual-physical connection is apparent in the Old Testament records as the nation of Israel repeatedly turned away from God toward material and self-centered desires such as power and wealth (a cause arising from the violation of spiritual law) only to find it once again being invaded and enslaved by its neighbors (an effect felt as the loss of God's favor and protection). The truth of spiritual laws can be verified based on their impact on individuals and society. Spiritual laws are subject to investigation just as physical laws are, but the unusually long time span between cause and effect can make it difficult to consciously make the connection. The investigative methods are necessarily different because the cause and effect relation of spiritual laws does not always manifest quickly or obviously in the material world, and the effect of sin, a violation of spiritual laws, may manifest years or lifetimes after the cause. Until Christians come to grips with the reality of soul reincarnation, they will be unable to properly understand the true relation between sin and sorrow in the world and will forever be mystified by the question "Why do bad things happen to good people?"

Scientists who are true to their profession use observation and experimental data to infer how the material world works, refine their ideas into theories, use mathematical and logical analysis to assess the theory, and formulate predictions that can be tested or may be able to be tested by future observations or measurements. Sometimes the analysis of an apparently unrelated collection of facts can lead to the discovery of a new potential truth that can be verified using observation and experimentation. Science is a self-correcting method for searching for the truth of the physical universe, and the scientific method encourages the modification, alteration, and discarding of a current theory if new evidence proves the theory wrong or defines a limitation beyond which the theory is invalid. Too often religious leaders use their interpretation of particular Bible passages to formulate what they consider or want to be scientific truth. They then work diligently to concoct a collection of unverified assumptions and compile a few false conclusions they need to support their version of truth. The process may include selecting a subset of physical observations and pseudoscientific conjectures to bolster their belief while ignoring or denigrating existing scientific research that disproves their conjecture. After all, if God has already granted you the truth through the infallible Bible, everything that contradicts it must be suspect, false, or an outright conspiracy to confuse the faithful and destroy their faith in God. They often use the scant scientific knowledge and outright scientific illiteracy of their congregations to manipulate the minds of the faithful into believing pseudoscience to support their religion and perhaps inflate their perceived self-worth and spiritual prowess among the faithful. Using the Bible to formulate pseudoscience does nothing to advance the spiritual knowledge of even the most devout Christian. It is difficult to have a conversation about science with someone to whom God has granted the gift of ultimate truth as reward for their faith and who demands absolute loyalty to that truth as a necessary condition to remain within God's inner circle and keep in his good graces.

Using a science curriculum to teach a child half-truths or twisted truth or outright lies because current scientific investigations do not conform to the church's interpretation of obscure Old Testament Bible passages does not serve churches or their members well. It is a method of deceit and indoctrination that speaks to the arrogance of certain Christian communities and their leaders and their disrespect of the hundreds of thousands of scientifically trained individuals who are able to seek greater knowledge of the material world while holding to a faith no less strong than the perpetrators of sham science textbooks. Yes, there are those in the scientific community who are not able to believe that a spiritual realm or God exists beyond their physical senses and scientific measuring apparatus. But these people are no different from the many nonscientists who come to the same conclusion for different reasons and the large number of former church members who have lost their faith because of the inappropriate stance and activity of certain churches, including the embrace of pseudoscience as religious doctrine.

Until a mind that has lost all awareness of its relationship with God finds and takes hold of something that stirs a response to the source of life deep within its consciousness, it will not begin to seek or see a spiritual meaning to life no matter what its rational beliefs. Depending on the individual, both science and religion can be a barrier to that awareness, but manipulating truth brings enlightenment to no one.

Attacking science or twisting its findings in a misguided attempt to protect religion just makes Christians look ignorant to many intelligent non-Christians and does nothing to serve the purposes of God or mission of Jesus the Christ, nor does it contribute to the spreading of the good news or expand the truth of the reality that God exists. Religion has no self-correcting mechanism to ensure the accuracy of its theology except perhaps meditation, which is not commonly practiced among Christians. New ideas about the relationship between man and God have historically been treated with contempt and fear. Contempt because the new ideas don't confirm or support the ideas historically or currently held as truth by religious authorities. Fear because those who hold to an accepted truth are often afraid that entertaining a different concept of truth will somehow endanger their souls or threaten their salvation and the salvation of those close to them. The good order of the church might be disrupted too, and heaven forbid that the authority and power of those in charge of the religious institution might be weakened or that members would leave the church and cause revenues to decline.

When Christians become a political force to protect their version of truth instead of a moral force for good without trying to impose their interpretation of good on others, they are attempting a top-down approach to spreading the message of Jesus. However, the good news of the Bible was never meant to be spread from the top down or Jesus would have gone to the religious authorities of his day and worked with them to convince them of his role as Messiah and encourage them to help him get his message to the people. Nor did he attempt to put hand-picked leaders into positions of religious or political power to support his ministry so that he could reach a wider audience more quickly or get laws passed that would allow legal enforcement of his moral code. He didn't try to change the existing religious and political structures or depose their upper hierarchy. His message was for those of the masses who were spiritually ready and mentally open to new ideas. Their lives would change, and that change would filter through the society and would ultimately cause change in the established religious and political organizations that influenced or controlled their daily lives. This method of spreading the gospel was put aside by Pope Sylvester I in 314 AD when he reversed the decision of the recently deceased Pope Miltiades and chose to cast his lot and the lot of the Catholic Church with the newly converted Emperor Constantine. He accepted material wealth and resources beyond his wildest imagination from the emperor in return for blessing and supporting the emperor's military campaigns as a better and more efficient means of spreading Christianity. By the time Pope Sylvester I died in 336 AD, the Roman Church had expanded from widely scattered small churches and chapels into a massive enclave on Vatican Hill filled with palaces, basilicas, and residences and all of the trappings of pomp and power (Martin, 1981). Apparently Pope Sylvester I decided Jesus didn't understand how to spread the truth of the gospel, and many other Christian leaders have come to the same conclusion and have fallen into the same trap as Pope Sylvester I.

When Jesus came into the earth, the existing forces of religion and politics undertook the task of silencing him to protect their authority and power. When he stood before Pilate to be judged to facilitate the desire of the Jewish religious authorities to see him killed, the following exchange of words took place.

> *"You are a king, then!" said Pilate. Jesus answered, "You are right in saying I am a king. In fact, for this reason I was born, and for this I came into the world, to testify to the truth. Everyone on the side of truth listens to me." "What is truth?" Pilate asked. With this he went out again to the Jews and said, "I find no basis for a charge against him." (John 18:37-38)*

When Jesus responded, "I came to testify to the truth," he meant that he was testifying to the Truth of God, the truth that God exists, the truth of his presence in the world as God's emissary and man's Messiah, and the truth of the way to live a lifestyle that would bring consciousness of God's presence to all who would follow his example. Truth is an aspect of the mind, and absolute Truth is a quality and attribute of the Mind of God. Jesus was truth incarnate in the sense that his mind was fully attuned to the Mind of God, and his understanding of the nature and purpose of God and humanity was complete and correct. His truth was the Truth of God, not truth as man incorrectly perceives it from the limited perspective of his conscious mind. Why didn't Jesus answer when Pilate asked him, "What is truth?" Perhaps it was because our perception of truth is constrained by our current state of knowledge and conscious experiences, and no one, not even Jesus, can reveal the truth to a person whose mind is not open to receiving new knowledge or accepting the truth. Pilate wasn't ready to hear truth and had no conscious or subconscious desire to let truth disrupt his current view of life and interfere with his position of power and authority. Even Jesus could not force an unreceptive mind to listen to truth, but of course Jesus never attempted to force anyone to believe his truth.

The final and absolute truth about the nature of the physical world and man's relationship with God resides with God himself. He knows when we were made, why we were made, why we now exist in human form living on a small planet in an immense universe, and why we seem to have so much trouble in our personal relationships within our families, within the societies we form, and among the various nations we create. God is Truth with a capital T, the ultimate source of knowledge because he is the master craftsman who created us and the world in which we live. He knows the how, the why, and the wherefore of our existence and destiny inside out and is willing to share all of that knowledge with us if we will seek him with all of our mind, body, and soul.

For truth in any clime is ever the same - it is law. And love is law, law is love. Love is God, God is Love. It is the universal consciousness, the desire for harmonious expressions for the good of all, that is the heritage in man, if there is the acceptance of the way and manner such may be applied, first in the spiritual purpose and then in the mental application, and the material success will be pleasing to any. [Edgar Cayce reading 3350-1]

Truth to the consciousness of man springs from the region where soul and God meet and are interwoven into one unified being and flows into the soul mind as the soul mind opens itself to its creator and is receptive to the knowledge God wants to share with it. Only then can the truth of God, the truth that God holds in his mind, move into and through the conscious mind to be manifested and experienced in the world of men and women. So, how does a scientist come to the truth of God? For that matter, how does a Christian or a practitioner of any religion come to the truth of God? It does not come from faithfully following the dictates, dogmas, and doctrines of other men who only think they know the truth or partly know the truth, even though their ideas may be helpful. Primarily it comes from sincerely seeking God and being willing to live a life that honors and respects that quest. Jesus lived that kind of life and set out the general principles for us to follow and master. He taught, demonstrated, and confirmed the way of life that would lead mankind to truth and eternal life. It is up to us to apply those general rules to specific life situations in a consistent manner through proper choices.

Truth emerges from Law,
Law arises from Love,
Love emanates from God.

The difficulty with ascertaining truth through religion is that religious truth comes primarily through revelation to fallible men and women who are not fully attuned to God. Whether it comes from putting your face in a hat and staring at a peep stone or from spending years in periods of prayer and deep meditation, when the revelator declares truth has been revealed to him, his audience pretty much has to accept or reject it without the opportunity for verification from independent sources. That makes it very difficult to separate the charlatans from the real prophets or even from the Son of God, especially when the message is new and first announced to the world. Jesus alluded to this difficulty when he told his disciples that the very elect will be deceived (Matthew 24:24) and made it clear that the deceivers will be able to accomplish their nefarious missions partly because they will gain the ability to dazzle their audience with feats that will astound them. This might be fake healings or stirring and emotional stories that pull on the heart strings; anything that lures a sincere seeker into trusting the would-be prophet instead of directly seeking God's inner prompting voice or remaining focused on the commandments of Jesus. A flair for oratory and the gift of a flamboyant preaching style are not signs that an individual or a particular denomination has special access to truth.

Those who compile and disseminate religious truth can have the same tendency to selfishness as everyone else and the truth they disseminate is only real and valid to the degree that they are attuned to the presence of God and allowing God to work through them. The Book of Revelation, received by the Apostle John during meditation, barely made it into the category of canonical text to become a book of the Bible. The book is highly symbolic and difficult to interpret, which gives religious authorities, laymen, and religious fanatics alike extensive leeway to imagine and formulate all manner of wild end-of-the-world scenarios from its pages. Even the strangest interpretations take on the mantle of truth when repeated often in the presence of a willing and receptive congregation. The readings interpret the same words of the Book of Revelation in a far different manner than fundamental and mainstream Christians, viewing the book as a highly symbolic description of the interface between the soul and the human body and the changes that the bodies and minds of humankind will undergo as their quest for a closer relationship with God leads to spiritual growth (Van Auken, 2015).

Revelation, like scientific theory, needs to be tested for accuracy and for truth. It is equally important to assess the sincerity of the one who declares he has been chosen of God to reveal his word to determine whether there is a component of ego and selfishness and the need to feel or exert power over others. There are at least two ways to test revelation. The first is for the seeker to let God bear witness to the words that are being spoken as revelation, that is, listen to the voice of God from within the inner consciousness as to the truthfulness of the message and the trustworthiness of the source of the message.

> *Listen to the voice within thee. For, His spirit will bear witness with thy spirit and guide thee in the way that thou shouldst go. [Edgar Cayce reading 262-59]*

> *… be not deceived by any. For as hath been given of old, it is not who or what will bring thee a message from above, or who from over the seas may present thee with a formula by which ye may use the influences about you to the greater application of materialization of forces. For lo, it is within thine own self. [Edgar Cayce reading 1992-1]*

The implied caveat with this method is that the individual seeking God's help to discern the truth must be receptive to God's presence by intentionally attuning his or her mind to the Mind of God, usually through meditation. The seeker must be willing to listen to the voice of God while filtering out interference from the self-serving voice of a potentially egotistical and power-hungry personality to recognize when corruption has entered the message. It may be particularly difficult to stay out of the influence of a charismatic personality who claims special revelation from God if one is susceptible to unhealthy dependent relationships. Too often the seeker is so desirous of being part of the new revelation that he is willing to set aside his will and mind and let them be replaced by the will and mind of a deceptive or misguided prophet looking for personal gain and glory, thus losing the ability to truly discern the spirit of the message and becoming one of the elect who are deceived. At its best, failure to correctly discern the truth of the revelation can lead to a group of people standing in a circle watching each other as the clock ticks off the last minute of the world, then another minute, and another minute, like the followers of the Millerism movement in mid-nineteenth century America. At its worst, it can lead to situations like the Jonestown mass suicides in Guyana, South America.

The second method of verifying the accuracy of revelation is to stand back and watch the fruits that it produces as people accept and act on it and apply the nitty-gritty aspects of the message to their everyday lives. If the message causes people to move closer toward a lifestyle that manifests the two commandments of Jesus and follows the example provided by the life of Jesus, that is a good sign. If the lifestyle of the new leader and the members who congregate around the message bearer becomes isolationist, divisive, and antagonistic toward nonmembers and outsiders, it is not of God. God does not seek division among his children; he seeks unity, and his word will unify those who adhere to and respect it. Too often though, different people will hear the same words and come to different conclusions about their meaning. This arises because of the different attunement of individuals to the presence of God within and different abilities to correctly discern the truth of the words. If the source of revelation tries to make the message more about him, more about money or material goods, or more about requiring specific beliefs and rituals than about loving God and serving neighbor, he shows himself to be a deceiver and should be avoided at all cost.

Revelations promoted as religious truth should be and can be tested. No one in any church should consider it heresy to question its religious tenets and doctrines nor should they be condemned by the religious authorities for bringing up doctrinal questions. Any church authority that would use the term "heretic" to chastise a member or a seeker or use the word in an attempt to force an individual to comply with the accepted revealed doctrine exposes himself as a false prophet. A sincere seeker does not have to agree with the doctrines and creeds of any prophet or religion to have a relationship with God. A sincere willingness to prevent material goods or worldly power from replacing

God, dedication to a regime of prayer and meditation to open our mind to the presence of God, and intentional action aimed at changing the focus and direction of our life toward honoring and loving God are far more important. Treating neighbor (that means friends and enemies alike) with respect and love will lead to a better understanding of God and a closer relationship with him than parroting the accepted doctrine. None of this requires a new religion or change of religion, although it may require a change in one's heart and belief system and will require an earnest desire to know the truth that sets men free.

> *And ye shall know the truth, and the truth shall make you free. (John 8:32)*

> *SELFISHNESS! Allowed, yes, of the Father. For, as given, He has not willed that the souls should perish but that we each should know the truth - and the truth would make us free. Of what? Selfishness! [Edgar Cayce reading 262-114]*

When we are asked to seek God within ourselves, it should not be understood to mean only a personal inner quest through prayer because that action alone can lead individuals to various concepts of truth. Truth is not fractured or splintered. The Truth of God is a single coherent set of facts that describe the nature of God as creator, his relationship to us, and his relationship with the material universe and spiritual realm. Different people come to different conclusions about the truth of God and his creations because we use our conscious minds to reason about him instead of seeking through the lens of our soul mind, our subconscious mind, to understand him. Prayer is an activity of the conscious mind; meditation is an activity of the subconscious mind. What we are missing if we only seek God through prayer and meditation is the power of activity based on faith and application of the fruits of the spirit to transform our minds and attitudes through the desire to be of service to others. Service allows God to express his generous love through us. This service activity is not only beneficial to the individual receiving the aid or comfort but strengthens the will and establishes and reinforces the correct spiritual pattern of mind in the individual who desires to be used by God as a channel of blessing to others.

This is a real transformative process within the soul that serves to bring the soul into closer attunement with God and becomes manifested in the outward behavior of the physical individual. As we express love in this manner, we begin to better understand the truth of God, not as an intellectual exercise but as an inner awareness, a form of selfless awareness where God is revealed to us as we imitate Jesus. The truth of his presence and nature will bathe our consciousness like the sun's radiation striking our bodies on a warm spring day and will usher in a sense of spiritual renewal. None of this truth, this new and greater awareness, has anything to do with religious dogma or doctrine. These things will seem trite by comparison. This fuller, more complete universal Truth from God, the single universal source of knowledge and truth, will flood the subconscious mind of each soul and flow into the conscious mind of each person, ready to be manifested in the world unimpeded by the rational aspect of the physical consciousness.

Souls – Portions Of God

Hence as He moved, souls - portions of Himself - came into being. [Edgar Cayce reading 263-13]

The first existence [of a soul], of course, was in the MIND of the Creator, as all souls became a part of the creation. As to time, this would be in the beginning. When was the beginning? First consciousness! [Edgar Cayce reading 2925-1]

What is a soul? The standard dictionary definition of soul is "the spiritual or immaterial part of a human being or animal." From the readings and biblical perspectives, this definition is weak, partly incorrect, and falls short of describing the true nature of a soul. The soul is not a component of the physical human body. A healthy human body is a biological organism made up of cells organized into flesh, muscles, blood, nerves, and other bodily components functioning in harmony as a single organism. A soul is a spiritual being that exists and lives in a nonphysical realm independent of any earthly body but which can and often does temporarily control a human body, being the source of consciousness and animation of that body. A soul manifests itself in the material world through the body, and the actions and words of the body continually express and manifest the soul's concept and understanding of God and its spiritual attunement to its Creator. The soul encompasses much more of creation than atomic matter and is eternal, not transient. The fact that we can love others, have hope in the midst of seemingly hopeless situations, have compassion for our fellow man, and can demonstrate many other noble moral and spiritual attributes is because our souls can manifest the higher spiritual and moral qualities of God through our bodies.

Why would God wish to create a soul? He already had angels to attend to him in the spiritual realm, to cater to his every whim and fulfill his every command. Angels already existed as created spirit forms before God sought the companionship of created souls. Angels were made to serve God in his creative efforts. They may be able to express delight and joy toward the marvelous creations of God, but do not have the ability to exercise independent creative powers and cannot share in the process of creation. Angels cannot choose to remain faithful to God; they are faithful and serving because it is ingrained within their nature. It is part of who they are and why they were created, but their service and loyalty is not a substitute for love.

God can only receive unconditional love from beings capable of giving love and who willingly choose to give their love to him. God is Love, and for love to be an active force, it requires an object toward which it can be directed and which is capable of returning that love. God's oneness and aloneness ultimately brought him to a state of overwhelming desire for loving companionship and led him to eventually create sentient beings with which he could share his creative abilities and the joy and wonder of his creations and toward which he could direct and express his infinite love. We call the beings that God chose to create for companionship souls. We human beings, in our limited conscious awareness, often use that term when we attempt to describe our innermost selves, not realizing that the soul is not a hidden esoteric quasi-physical component of the body. This confusion arises because in our current state, we tend to think we are a physical body that possesses an inner spiritual core. Instead, each of us is a soul that currently inhabits and directs a biological physical body (805-4 and 5749-14). Most of us can no longer imagine or comprehend our true nature as a soul because we suppress our soul consciousness and allow our physical consciousness to lead us while we are incarnated in the physical realm.

> *All souls were made as companions, as a portion of the Whole.*
> *… For it is in the image, in the pattern of the Creative Forces;*
> *and only that as is constructive can abide in that Creative*
> *Presence. [Edgar Cayce reading 1230-1]*

The soul is an integrated unit of spirit activated by a mind under the direction of an independent will. Souls were created from spirit, the eternal substance that constitutes God and is the essence of God. The pattern for the created souls was none other than God himself. Each soul was given life as an extension of God, a quasi-independent portion of the all-pervasive Spirit that is God, while forming and retaining within itself a separate identity and consciousness. Souls were not just given a portion of God's Spirit; instead, and more accurately, each soul is a portion of the Spirit that is God. This basic unit of God spirit was given independence in the form of a mind that is able to direct and shape spirit into manifested activity and a will with which to guide and direct the mind into desired channels of thought. There is nothing that is not God, and that includes our souls, our individual essence, our individual "I AM."

Souls have form, often called the soul body when active in its natural nonphysical residence, and sometimes referred to as the astral body when in association with a physical body (900-348 and 5756-4). Souls are not something that we humans possess; souls are that which we are, our true selves. Although souls are created beings and had a beginning when they came into existence in the spiritual realm, they are potentially eternal and immortal. Their eternal nature and immortality is derivative to their being a portion of God, not from any other property with which they are endowed and they will remain in that holy state only if they remain consciously aware of their relationship with God and hold true to the purpose for which God created them. Their purpose is to be loving companions to their Creator with the ability to share in creative activities and the joy and awe of his creative endeavors.

In His Image

God created souls in his image. The activity of soul creation followed the same spiritual law that is reflected in biological law, that "like creates like." A spirit being (God) created a spirit being (soul) out of himself and in his likeness in the spiritual realm with the potential to be like him, just as human beings in flesh (human parents) create a flesh infant being (human child) in their likeness in the physical realm with the potential to be like them. We see the biological expression of this law everywhere in the natural world. A plant produces a seed, which brings forth another plant of the same or similar type when the proper conditions of temperature and moisture that will sustain the new plant are met. It is incorrect to think of souls as being composed of any of the physical elements that make up this material universe. Souls are composed of spirit, the same substance that is God, and primarily are active within a nonphysical framework defined by a set of laws similar to the physical laws that apply in the world of matter and energy. Spiritual laws also define the metes and bounds God has set on the conscious activity of the soul and have their physical counterpart in moral codes of human behavior. These spiritual laws apply to souls even when they are incarnated into physical bodies that are themselves subject to the physical laws of the universe.

Only beings created in His image have divine attributes and characteristics and the potential to become his true and loving companions. Of all creation, the soul is the greatest because it alone has the inherent capacity and potential to be in a state of true loving companionship with God, to be a companion to him in his spiritual realm, and to participate and share in his creative acts. Like God, the essence of each soul is spirit. Like God, each soul has an independent mind that is used to guide and mold spirit into manifestation, and like God, each soul has a will that can be used to direct and steer the mind into desired creative activities. Our souls are literally created from Spirit, from God himself, who gave of himself that we might have our own independent existence with the ability to cocreate with him by focusing spiritual energy through the activity of the mind.

Each soul is endowed with a mind. The mind is the active force within God that allows him to direct and shape spirit, to create in the spiritual realm, to create the physical universe, and to create life within the physical world. There is only one mind associated with a soul and that forms the individuality of the soul, but it is convenient to imagine the mind of the disincarnate soul as having two components: the subconscious mind that serves as the primary seat of awareness of the soul and the super conscious mind that serves as the interface between the soul and God. When a soul incarnates into a human body, the mind forms a third aspect, the physical consciousness that acts as the interface between the soul and the brain and sensory system of the body (Figure 1). The awareness of human beings is primarily at the physical consciousness level, the subconscious mind of the soul being suppressed and relegated to making its presence known only through the control of certain autonomic functions of the body and as dreams during diurnal periods of sleep. When the soul is not incarnated in a physical body, the physical consciousness of the mind is not present, and what we humans call the subconscious mind becomes the center of conscious awareness of the soul, the soul mind.

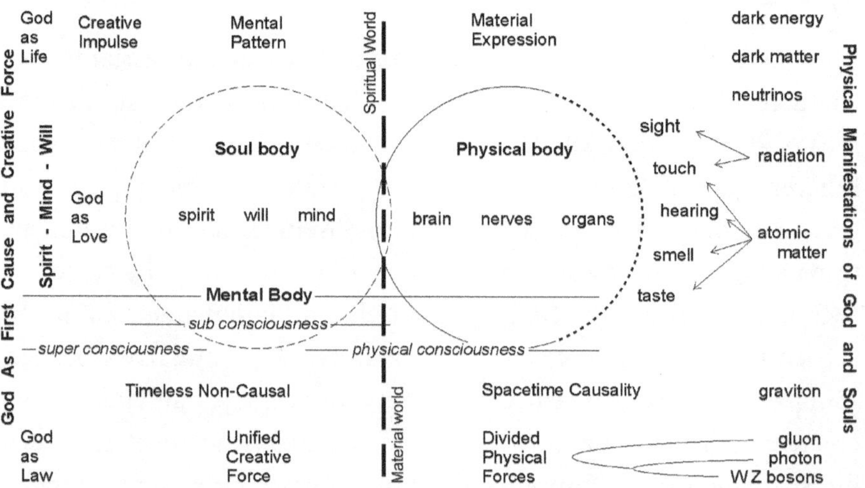

Figure 1. The soul and physical body in relation to spiritual and physical realms.

The mind is the creative force within a soul, the aspect of the soul that uses spirit as raw substance and manifests in the spiritual and physical realms in response to the directives of the will and in accord with the soul's comprehension and understanding of its God. The mind controls the actions of soul bodies, our ethereal nonphysical bodies that reside in the nonphysical realm. The mind controls the actions of physical bodies occupied by incarnated souls in ways that illuminate each individual's concept of God and willingness to remain true to the purposes of God through material activity. When the mind is trained to adhere to God consciousness, it naturally manifests its innate divine spiritual qualities through personal relationships with other incarnate souls. When the mind is left unbridled and allowed to habituate to and focus on selfish interests at the expense of its God consciousness, as is the wont of too many individuals in the material world, it manifests the forces of evil through self-centered and self-gratifying motives that denigrate others or subject the bodies and minds of others to the personality of the selfish individual. Evil is a misuse of spirit, a self-directed, soul-directed mental distortion and misapplication of spirit that is manifested when a soul rebels against its creator.

> *Spirit is the First Cause. Mind is an effect, or an active force that partakes of spiritual as well as material import. Mind is an essence or a flow between Spirit and that which is made manifest materially. [Edgar Cayce reading 262-123]*

In the beginning, the soul mind was a true likeness, a true image, of the Mind of God, and all souls were conscious and mindful of the presence of God. Each newly minted soul, each independent soul mind, was attuned to the Mind of God and had not as yet become altered, defiled, corrupted, or distorted by the yearning that arises from mental desire. This God consciousness was facilitated by the super conscious mind, that part of the mental body that is the bridge and life-sustaining link between the created soul and its Creator. This does not mean that every soul was privy to the thoughts of God, but it does indicate the incredibly close relationship between the soul and God. Similarly, we humans are rarely aware of the thoughts and concepts present in our subconscious mind even though it is a higher aspect of the same physical consciousness that we use every day to guide our physical actions and use to reason and make rational decisions.

In this original state of the soul at the beginning of our independent existence from God, our soul minds were aware of God's plan and purpose for us; that each soul was created to be a companion to God and a co-creator with God. The original, unspoiled soul mind was impressed with a mental and spiritual pattern that allowed it to be a full member of God's spiritual kingdom. God knew that the original foundation of this relationship was precarious. The gift of independent thought and activity he granted to the soul could lead it to fulfill its purpose through an increased awareness and knowledge of God and closer mental association with God or could lead it to selfish activity and descent into a state of lost God consciousness and mental separation from God.

> *What, then, is WILL? That which makes for the dividing line between the finite and the infinite, the divine and the wholly human, the carnal and the spiritual. For the WILL may be made one WITH HIM, or for self alone. With the Will, then, does man destine in the activities of a material experience how he shall make for the relationships with Truth. [Edgar Cayce reading 262-81]*

Will is the impulse that directs and guides the mind to express and glorify God, or to express the selfish interests of the soul and glorify itself. Will has the ability to create rebellion, chaos, and separation from God in the spiritual and physical realms and the strength to reshape the mind of a wayward soul and bring it back toward the original holy and sanctified state it occupied as befitting a spiritual being created in the image of God. When the will of the soul seeks cooperation with the will of its Creator, the mind and purpose of God for that soul will be recognized by the soul mind. The Mind of God will become an active cooperative force within the soul mind, and God's divine purpose for us will be manifested through the physical body being animated by the soul. When the will becomes enamored with or attached to desires that contravene the natural attributes and expressions of God, events are set in motion that cause the soul to lose conscious awareness of God. The original mental and spiritual pattern that brings awareness of the Creator still resides within the soul mind, but access to it becomes clouded and blocked by repeated and habitual attachment to selfish thoughts and desires.

Each soul was endowed with will and the freedom to exercise it in accordance with or in discordance with the plan and purpose that God intended for it. The soul could never exist as an independent consciousness separate from God unless it had the faculty of will under its own control. It would only have been an extension of the consciousness of God. But true companionship between this independent entity and God was only possible if the soul was willing to remain a true image of God without force or coercion. The will of each soul is individual and is free to act on desires that arise in the mind. Those desires can be consistent with the will of God or can be self-centered, self-gratifying, and counter to the will of God. The only manner by which a soul may separate itself from God and possibly lose its eternal nature is through its own improper exercise of will. Each soul was (and still is) repeatedly given the opportunity to make choices consistent with its innate desire for companionship with God and desire to remain attuned to and true to the plan and purpose of God. That opportunity is still available to every soul, disincarnate or incarnate. God is an omnipotent and omnipresent being, but he does not know if the soul will make choices consistent with his plan or counter to his plan until the soul makes and acts on a particular choice. God does not violate our free will by forcing us to make the choices he prefers.

> *He did not prevent [the fall of man], once having given*
> *free will. For, He made the individual entities or souls in*
> *the beginning. For, the beginnings of sin, of course, were in*
> *seeking expression of themselves outside of the plan or the*

way in which God had expressed same. [Edgar Cayce reading 5749-14]

The Maturing Soul

An analogy to the original relationship between the soul and its creator is the relationship between a human infant and its mother. An infant has free will but limited physical, intellectual, and emotional experiences that it can access and its previous existence and history as a soul is beyond reach of its newly established and untrained physical consciousness. Without knowledge of these experiences, the free will of the infant cannot be fully exercised, and its independent self-awareness exists mainly as an unrealized potential. Similarly, the mind of each newly created soul also contained all of the soul's potential to form or shape the identity and individuality of the soul, unrealized potential because there was as yet no mental desires upon which the will could take hold and direct the mind to dwell upon and activate. This initial state of the soul was somewhat analogous to the tabula rasa, or blank slate, mind of a human infant, full of possibilities not yet identified or formulated. The infant soul mind resonated with the Mind of God and was intuitively aware of its oneness with God, but at this stage of development the soul consciousness was mainly derived from the super conscious mind, the portion of its mental body that linked to and interfaced with the Mind of God. The soul mind was innately attuned to the Mind of God, the Creative Force that brought it into existence.

The soul was aware of God's existence and presence and held within itself the fundamental characteristics and qualities of God. The original impulse of a soul when it was created as a spiritual being was to love and glorify its creator through its experiences and activities. The soul was innately aware of its relationship with God and may possibly have understood the incorruptible righteousness of God. The soul was also curious and perhaps interacted with its environment with a sense of bewilderment and fascination. As the soul mind, the subconscious mind, began to develop and become consciously aware of its independent existence and its surroundings, the will of the soul was presented with opportunities to act on desires that could maintain and strengthen a loving relationship with God or could develop and enhance a sense of self-centeredness and selfishness. The more it explored selfish desires, the farther it wandered away from God in consciousness. As the mind began to habitually satisfy selfish desires it eventually created a formidable mental barrier that interfered with the soul's awareness of its creator.

Maturing souls were like maturing humans. Like many human parents, God created loving, sweet children who grew up to be teenagers, those too often self-centered beings who know so much more than their parents and live in a perpetual state of rebellion. Teenagers know far more than their adult parents when it comes to deciding what is exciting and fun to do and what is best for them. They often don't comprehend or take the time to think about the repercussions of their behavior or the pain and suffering it can inflict on their health and wellbeing or that of their parents and the larger society in which they live. As the level of soul awareness increased to where souls were able to mentally explore their creative abilities, many of them made poor choices that contributed to a gradual separation from God through loss of God consciousness and loss of awareness of their origin. Although teenage souls probably couldn't bring premature gray hairs to God's head, they did bring him exasperation (Genesis 6:6) and concern for their future. Despite the anguish their behavior inflicted upon God, he still desired to help them see the error of their ways. He wanted and needed for them to change their behavior for the better, not just because he desired their companionship but because their behavior was self-destructive and created disharmony in the unity that is God.

When the will of the soul is repeatedly used to encourage the mind of the soul to dwell on a certain pattern of behavior or to set specific desires into motion, the soul mind is altered. When the desires are selfish, that is, are self-serving to the soul without regard to the soul's relationship to God or fellow souls, the soul mind shifts out of resonance with the Mind of God. The soul loses awareness of its creator as it chases selfish but satisfying mental diversions and assumes a higher sense of self-worth than its true value. These thoughts and actions in spirit began to create dissonance and discord in the spiritual realm. This spiritual rebellion is alluded to in the Bible by the fall of Lucifer, the morning star, for exalting himself above God (Isaiah 14:12 and 262-89). This is not a reference to the banishing of some powerful spiritual being or an evil twin of God called Satan but is a reference to the fall from grace of souls like us. There were many rebellious souls who were more than willing to strike out on their own without thought of the consequences of their actions because in their minds they no longer needed God.

The power of the will to lure the soul away from its creator for the fulfillment of the soul's own selfish desires and work in opposition to God's will for the soul is often symbolized in the Bible by Satan or evil. Lucifer may or may not have been an individual errant soul but Lucifer as used in the Bible is an archetypal representation of the rebellion of souls against God.

> *As has been given, error or separation began before there appeared what we know as the Earth, the Heavens; or before Space was manifested. [Edgar Cayce reading 262-115]*

The will of a soul has the power and freedom to direct its mind to remain faithful to God's plan for the soul, to become and remain a true companion, by acting on mental desires that are consistent with the will of God. The soul mind is able to direct spirit into activities that are virtuous, righteous, and consistent with the inherent goodness that is part of the nature of God and the souls he created. That was the original intent when souls were created. The mature soul that willingly remains in a state of mental attunement with God understands its relationship with God and realizes that the only thing that can cause it to separate from God is its own inappropriate and willful corruption of that relationship. This is what God wants from us. An unwilling companion is not a true companion, only a slave having no choice but to express a false sense of companionship out of fear of the consequences of expressing its true thoughts, or an automaton that expresses companionship because it is hardwired to only behave in that manner.

Freedom is a two-edged sword. The free will God granted souls gave them the capability to turn away from God, which many of them did long before the universe was created and souls entered the earth environment. When God created souls, he was aware of the possibility and probability that some of them would express their independence by becoming enamored with their own self-centered desires and exciting creative abilities and would turn away from him. God could not grant freedom of expression to souls without opening the door to the possibility that they would misuse that freedom and descend into selfish and evil behavior that defiles and demeans the goodness and love that he represents. The accumulation of mental and emotional experiences in the spiritual and physical realms does not automatically confer wisdom to the soul mind. Even mature souls who have had a long series of incarnations in human bodies can make poor life choices because they have refused to align their wills with the will of God. They have not learned that it is far more rewarding to use their free wills to choose eternal life over the fruitless quest for seductive carnal pleasures and ego-driven power trips.

Many souls remained faithful to God and to the purpose for which they were created. Their desires remained unselfish, and their wills directed and changed their minds in ways that developed their individuality but retained the qualities and attributes that are the essence of God's nature. Other souls chose to squander their inheritance through selfish thoughts and activity. They sacrificed their original loving relationship with God for the pleasure and excitement they derived from selfish desires. Exactly what form these rebellious desires can take in the spiritual realm is difficult for the physical conscious mind to imagine, but they involved aspects of soul expression that were inconsistent with God's plan and purpose for souls. Souls may have begun using their will in a selfish manner to explore and experience aspects of their environment and to create in ways that were not in accord with the higher purpose that God had hoped would guide each soul. Perhaps souls can experience sensations or stimuli in the spiritual realm that are akin to carnal desires in the earth; desires that draw their attention away from God, cloud their awareness of God's presence and love, and even block the flow of God's love into the soul mind.

> *It is true that the activities so far as in this sphere or galaxy of activities of the planetary forces within this present solar system, the earth first became as the indwelling of the consciousness of the race or the man in this particular sphere, but sin - the separation - that as caused the separation of souls from the universal consciousness - came not in the sphere of materiality first, but in that of spirit. [Edgar Cayce reading 1602-3]*

It is possible that jealousy, jockeying for favoritism from God, trying to become better than other souls, or holding and displaying other unloving attitudes caused the downfall of many souls. Some souls may have found it convenient or desirable to gain advantage over their fellow souls or perhaps reacted with anger toward other souls over some perceived slight or difference of opinion. There is no reason to expect that such conduct as jealousy, greed, and anger is limited to human relationships in a physical world. There is no reason to believe that souls cannot experience emotion and react with many of the same unsavory characteristics that are recognizable in human behavior, such as the desire to take advantage of another person or gain at another person's expense or react with hatred when one feels slighted by the actions of another. In fact, human behavior is the material expression of the thoughts and desires of an incarnated soul, not something that arises from inside the brain of a human body. The subconscious mind of the soul is united with the human conscious mind and its thoughts guide and direct human activity. The readings describe the first soul incursions into the earth as thought forms designed to experience carnal gratification even before he had prepared a suitable life form for soul education through physical incarnation. This indicates that souls have the ability to move freely through different realms, perhaps driven by the pursuit of pleasure and other selfish motives.

ALL force, all power that is manifested in the earth, EMANATES from a spiritual or God-force - as it manifests. MAN in his madness, or in his selfishness, TURNS same into that which becomes either as miracles in the experiences of man or crimes that make for the crying out of the people who heed not. Know that the Lord IS GOOD! [Edgar Cayce reading 815-3]

Souls operate within a framework defined by a set of spiritual laws, similar to the manner in which the physical world of matter and energy operates within a set of physical laws. There are spiritual consequences when a soul decides to act in opposition to God's plan for that soul. Thought and activity by souls in the spiritual realm can be just as destructive as words and activity by soul-directed humans in the physical world. God as spiritual law has prescribed the set of rules that makes souls accountable for their thoughts and actions. God's spiritual laws influence and confine every activity of the soul and define the arena within which a soul can move and can express itself but does not dictate any soul's response to events in the spiritual or physical realms. Spiritual laws were established to protect and educate the soul. Spiritual laws do not violate our free will but do define the consequences for each choice we make using our free will. One significant difference between the material world and the spiritual world is that the spiritual world is not causal, or at least there is no conscious sense of an ordered sequence of events in time. Souls in the spiritual realm are not always aware that their choices and decisions are detrimental to their continued independence and their existence in a state of oneness with God. It seems to defy logic to say souls are not aware of the consequences of their actions in the spiritual realm but there is no reason to expect the logic of the physical mind to apply to the subconscious mind and the spiritual realm.

The same spiritual laws that operate when souls are disincarnate in the nonphysical realm also operate when souls are incarnate in physical bodies and work in conjunction with the physical laws of the material universe that operate on the material body. The growth and development of the soul mind, like the growth and development of a human body, is dependent upon the food it receives and assimilates. The ability for a soul to achieve its full potential as a true companion of God can only be realized if it is fed the proper mental nourishment. Unfortunately, the food that it craves and too often ingests comes in the form of a steady diet of selfish and carnal gratification that the will identifies as pleasurable. The will can reinforce the pleasurable feeling of this spiritual equivalent of a sugar high by repeatedly allowing the mind to sample and experience the thought pattern and physical activity associated with the tempting but harmful carnal experience. It is imperative that the will make the decision to turn the soul mind away from pursuit of these detrimental experiences and toward the higher ideals that can cleanse the soul.

> *As the soul seeks, then, for that which is the sustenance of the body - as what the food is to a developing, a growing body, so are the words of truth (which are life, which are love, which are God) sought that make for growth, even as the digesting of the material things in a body make for a growth. This growth may not be felt in the consciousness of materialization. It is experienced by the consciousness of the soul, by which it enables the soul to use the attributes of the soul's food, even as the growth of the body makes for the use of the muscular forces or attributes of the physical body. [Edgar Cayce reading 254-68]*

Religion and the Soul

During the first few hundred years of its existence, the early Catholic Church spent a considerable amount of energy and effort to solidify and codify the "correct" religious doctrine that it expected its members to wholeheartedly believe. The concept and nature of the soul became entangled in the theological bickering of the early church and political infighting between the Roman Catholic pope and Emperor Justinian of Constantinople. In an effort to discredit the theological doctrines of Origen and other early writers that did not adhere to the theological inclinations of Justinian, he convened a quasi-legal ecumenical council (Fifth Ecumenical Council or Second Council of Constantinople) in 553 AD to counter these so-called heretical doctrines and exert his authority over the doctrines of the early church. He also used the council as part of an attempt to unify his empire by compelling religious uniformity in his subjects and to draw political and religious power from Rome to Constantinople. The Roman Pope Vigilius finally, albeit reluctantly and despite the antagonism he felt toward Emperor Justinian, wrote a letter to the Eastern Orthodox Patriarch Eutychius wherein he accepted the capitulas and anathemas issued by the council on the nature of God, Jesus, the Trinity, and other religious matters. Although he ultimately agreed with and approved the decisions the council rendered, despite his misgivings about the jurisdiction of the council, later historians claim that he did not specifically agree to denounce the doctrines of Origen. In the midst of the interminable slicing and dicing of doctrine and attempt to understand the nature of Jesus and the Trinity through dogmatic nitpicking, the church smothered the knowledge and fact of the preexistence of the soul. No person of good standing in the faith was allowed to believe that any person's soul existed before its human body was formed. The church severely hindered the ability of religious scholars and laymen alike to understand the true nature of the souls that inhabit and animate the bodies of every human being. Future honest investigation into the true nature of Jesus and the full extent of his activities on the earth was crippled by the perceived need to impose enlightenment from the top down, mainly for political reasons.

Most conclusions of the council were issued as anathemas, formal curses that denounced a belief as forbidden, and anyone who adhered to that belief was declared heretical (Schaff and Wace, 1900). Many of the anathemas are complex theological statements written in such convoluted and obtuse language as to be nearly incomprehensible. Some anathemas are simple, straightforward statements of what a true believer must believe or cannot believe if they are to remain in the church. There is no reason to think that any one of these many directives represent the absolute truth. They were designed to forge a common system of thought, to control any of the faithful who might dare to think differently and thus upset the good order of the church, and to enforce a tightly controlled sense of community in the people who did accept them. Everyone else was kicked out of the club. They kept the church doctrine pure by instilling the fear of excommunication and subsequent eternal damnation for those who would question it. It was spiritually and politically risky, and at times physically risky, to declare an opposing opinion or express a desire to further study the subject. The doctrinal pronouncements of the council can be grouped into the following categories: fifteen anathemas against Origen, nine anathemas of the Emperor Justinian against Origen, and fourteen capitulas of the Council (which are also anathemas). Two of these newly established doctrines are of importance herein:

Anathema 1 of the anathemas against Origen: If anyone asserts the fabulous pre-existence of souls, and shall assert the monstrous restoration which follows from it: let him be anathema.

Anathema 1 of the Emperor Justinian against Origen: If anyone says or thinks that the soul of the Lord preexisted and was united with God the Word before the Incarnation and Conception of the Virgin, let him be anathema.

When Paul said, "[Jesus] is the image of the invisible God, the firstborn of every creature" (Colossians 1:15) or "[Jesus] is the image of the invisible God, the firstborn over all creation" (Colossians 1:15 NIV), he was explaining that Jesus (actually, the soul of Jesus) was created before all biological life and even the physical universe (the Greek word *ktisis* translated as "creature" and "creation" in this passage means "the act of creating"). As the firstborn of every creature or creation, the birth of the soul of Jesus preceded that of any other human or animal life in the universe, preceded the creation of the universe itself, and preceded the creation of all other souls. The soul of Jesus was the firstborn of all created souls and, like all souls, was created in the image of God despite the assurances of Pope Vigilius and the Emperor Justinian that the soul of Jesus did not exist before his birth on the earth. The soul of Jesus and the souls of all men existed before their physical bodies were conceived or born. The word "image" used in this verse certainly does not mean that the man Jesus was created with a physical appearance that is somehow similar to the appearance of God. Only souls are created in the image of God.

> *In the beginning, celestial beings. We have first the Son, then the other sons or celestial beings that are given their force and power. [Edgar Cayce reading 262-52]*

No earthly physical body has ever been made in the image of God. Human bodies are simply vehicles that souls occupy to facilitate the search for their lost status as children of God. Everything that grows from the earth or is born on the earth has a beginning and will also have an ending, a finite life span and a physical death. Because souls are created by God from God, they bear the likeness of God and have qualities that are consistent with the nature of God. This is a spiritual as well as a physical law. On the earth we see this same law in that the fruit of an oak tree does not germinate and grow into a pear tree. No human body has ever inherited the qualities of God; God is the God of souls, and the bodies of men and women are the physical means by which souls express themselves in a causal world of matter and energy. Being the firstborn of all creation in the image of God implies birth in the spiritual realm and existence before physical birth, but religious authorities in the early church decided that any person who thought that souls could or did exist before the human body it occupied was worthy of the church's condemnation, and thereafter no church member in good standing dared suggest that Jesus might have had a soul that preexisted his body.

Souls have the same inherent nature as God. Jesus alluded to this true nature as well as his special relationship with God when he told the Jews "I and my Father are one" (John 10:30). When they denounced him for blasphemy and picked up stones to throw at him, he followed with the corollary, "Is it not written in your Law, I said, Ye are gods?" (John 10:34). Souls, created out of God in the image of God, have the potential to be gods, not the entirety of God or the full equivalent of God, but like God in that they have the capacity to create by using will and mind to shape and direct spirit. When souls use that will to create in ways that are consistent with God's purpose, they comply with spiritual laws, and the line of communication between God and soul through the super conscious mind remains open. When we decide we know better than God and use our will for selfish purposes and to conduct ourselves contrary to God's plan, we confront the law. Our thoughts and action set in motion a definite consequence, and in the process we interfere with or impede the lines of spiritual communication until we can barely detect God's presence and no longer recognize his activity within us or in the world around us.

> *The LORD brought me forth as the first of his works, before his deeds of old; I was appointed from eternity, from the beginning, before the world began. When there were no oceans, I was given birth, when there were no springs abounding with water; before the mountains were settled in place, before the hills, I was given birth, before he made the earth or its fields or any of the dust of the world. I was there when he set the heavens in place, when he marked out the horizon on the face of the deep, when he established the clouds above and fixed securely the fountains of the deep, when he gave the sea its boundary so the waters would not overstep his command, and when he marked out the foundations of the earth. Then I was the craftsman at his side. I was filled with delight day after day, rejoicing always in his presence, rejoicing in his whole world and delighting in mankind. Now then, my sons, listen to me; blessed are those who keep my ways. Listen to my*

instruction and be wise; do not ignore it. Blessed is the man who listens to me, watching daily at my doors, waiting at my doorway. For whoever finds me finds life and receives favor from the LORD. But whoever fails to find me harms himself; all who hate me love death. (Proverbs 8:22-36)

Perhaps better than any other passage in the Bible, these verses from Proverbs present a fundamental premise in the philosophy of the readings. They describe the preexistence of Jesus as the first soul created by God, his role in the creation of the future dwelling place of humankind, his stature as savior of humankind, and his promise that all who seek God by following his example and teachings will be reunited with God. Jesus becomes the instructor of humankind and the source of true life, a spiritual life that man can seek and find. Note that God is the creator, the Master Craftsman, and the soul of Jesus was the craftsman helper at God's side working with him to create the universe and establish the earth as a suitable habitat for humankind. Perhaps it is only fitting that the man Jesus was born into the household of a man who worked as a carpenter, a craftsman whose building material is wood, and for a while Jesus would help his earthly father in the carpenter trade.

Twin Souls and Soul Mates

At first glance, the readings might seem to suggest the concept of twin souls, whereby two bodies can be occupied by one soul or share a single, yet in some way divided, soul. There appear to be few readings that address this topic, and they do not seem to unequivocally address this question but do strongly suggest that the inference of twin souls is not correct in the sense of a soul division whereby one soul divides itself such that it is capable of animating two bodies instead of being limited to one body. Four relevant readings that touch upon the concept of twin souls are readings 3285-2, 5749-8, 5749-7, and 263-13.

That there are identical souls, no. No two leaves of a tree, no two blades of grass are the same. They are the complement one to another, yes; but these are dependent upon the purpose. [Edgar Cayce reading 3285-2]

In reading 3285-2, the source of the reading implies that the questioner has an incorrect concept of twin souls, but we don't really know her understanding of the concept of twin souls. The question actually has two components: whether the concept of twin souls is true and whether the questioner and her husband are twin souls, the latter component making the first component of the question less specific. The answer is that there are no two identical souls. It is possible that the answer addresses the question as to whether the husband and wife have identical souls and not the larger question as to the existence of twin souls in general, but the additional statement that no two objects in nature are the same suggests that the answer refers to the general case. It seems to imply that no one soul has divided itself to occupy the bodies of the husband and wife, and the concept of twin souls as the same soul occupying two bodies is incorrect. However, it does indicate that the idea of twin souls can be used to describe two souls when one soul is the complement of another soul in purpose, the reason behind each soul's incarnation. Two souls may be in a close physical relationship because there is a common spiritual and physical purpose for them to express within that relationship.

> *Is the teaching of the Roman Catholic Church that Mary was without original sin from the moment of her conception in the womb of Ann, correct? (A) It would be correct in ANY case. Correct more in this. For, as for the material teachings of that just referred to, you see: In the beginning Mary was the twin-soul of the Master in the entrance into the earth! [Edgar Cayce reading 5749-8]*

In reading 5749-8, the source initiates a discussion of the concept of a twin-soul in response to a question about the accuracy of the Catholic Church teaching "that Mary was without original sin from the moment of her conception in the womb of Ann." The words that Mary was the twin-soul of the Master "in the entrance into the earth" seems to refer to the common purpose of shared activity of two incarnated souls as associated with the new lives they are beginning on the earth. There are a couple of complementary ways to interpret the answer. It could refer to Mary's birth sixteen years before the birth of Jesus as a separate soul involved in the shared and common purpose of providing the means for the incarnation of Jesus. The words "in the beginning" might also refer to the entrance of the soul of Jesus into the earth as Adam and the entrance of the soul of Mary into the earth as Eve when both were devoted to the shared and common purpose of restoring God consciousness to souls that were lost to the attractions of materialism and self-indulgence.

> *And when man had reached that period of the full separation from Creative Forces in the spirit, then flesh as man knows it today became in material plane a reality. Then, the immaculate conception is the physical and mental so attuned to spirit as to be quickened by same. Hence the spirit, the soul of the Master then was brought into being through the accord of the Mother in materiality that ye know in the earth as conception. [Edgar Cayce reading 5749-7]*

Reading 5749-7 specifically talks about Jesus and Mary in the context of immaculate conception. It clearly states that the soul of Jesus was attracted to the future child of Mary because the mental desire and activity of Mary created that opportunity and the correct family environment (Imes, 2021). It doesn't mention or imply that the soul of Jesus was already active in the world through the body of Mary as a twinned soul. According to the readings, one of the main points of attachment between soul and body is the lyden gland, probably associated with the cells of Leydig. The body of Mary was animated by her soul for sixteen years before Jesus was born. At the birth of Jesus, did Mary's soul partly detach itself from her lyden gland and attach itself to the lyden gland of the infant Jesus? This idea seems extremely implausible and too much of a stretch. It could also be interpreted to mean that the soul of Jesus would materialize, begin to exist, be "brought into being" at the moment of conception, but we know that is an incorrect understanding. The readings clearly establish that souls existed before the universe was created and that soul birth, the attachment of the soul to a new body, occurs at or close to the time of the actual physical birth of an infant. The soul of Jesus existed before his birth in the physical world and was brought into the earth through incarnation, not brought into existence at conception.

> *Each soul choosing such a body at the time of its birth into material activity has its physical being controlled much by the environs of the individuals responsible for the physical entrance. Yet, the soul choosing such a body for a manifestation becomes responsible for that temple of the living God, when it has developed in body, in mind, so as to be controlled with intents, purposes and desires of the individual entity or soul. [Edgar Cayce reading 263-13]*

Reading 263-13 speaks more generally about soul birth or incarnation and discusses the fact that the incarnating soul is independent of the souls manifesting in the bodies of the parents and other family members. Although the soul usually chooses the body it is to inhabit, the parents control the environment and material circumstances that the soul will experience during its early years on the earth. The incoming soul is responsible for its own life choices after it has sufficiently developed in body and mind so that its own purposes and desires become the controlling factor in its life. The incarnating soul is not a shared spiritual entity with the souls of its parents or any other family member. It is fully independent in spirit and mind and creates its own destiny.

In summary, the readings do not clearly support the idea of a divided or twinned soul as a single soul that occupies two bodies or as a soul that projects part of itself having one quality into one body and part of itself having another quality into another body. The term "twin souls" as used in a question or answer in the readings does not mean that a soul has split or divided itself so as to simultaneously occupy two bodies active on the earth during some overlapping interval of years.

The word "twin" is used instead in the context of a shared purpose and desire and willingness of two souls to work together cooperatively in unison over many lifetimes. The souls of Jesus and Mary worked in tandem, each reinforcing and supporting the other, to accomplish a shared desire to restore the souls of men to their original and rightful place as children of God. They began their long journey together as Adam and Eve, and because they brought error (sin) to those incarnations, they were required to complete a full sequence of earthly lives until they were purged of that tendency.

The souls of Edgar Cayce and Gladys Davis, his assistant and recorder and compiler of the readings, are not described using the terminology of twin souls, but their shared activity throughout many lifetimes to bring enlightenment to the world qualifies them as twin souls in the sense of having a shared purpose. Their shared activity as narrated in reading 294-9 is a uniting for a common purpose "from the beginning," which could be taken to mean from shortly after soul creation, referring to the souls that remained faithful to God's purpose, or from the period of mass soul migration into Adamic bodies as part of an organized attempted to bring lost souls to a restored understanding of God. They were advised to be faithful to one another in their spiritual endeavor but warned to not be distracted by physical attraction. They were called "souls of the making", which suggests that they were on their own personal spiritual journey as well as incarnating for the benefit of others.

Whereas twinned souls seems to refer to two souls who share a common (higher?) purpose in the spiritual realm and who subsequently incarnate with the objective of working together on the earth to bring that purpose to life in materiality, soul mates as used in the readings seems to refer to two souls who have shared good experiences on the earth and as a result are drawn together because of those physical experiences and because of the prompting arising from soul memory. There are only a few readings that directly describe or mention soul mates. One reading (1556-2) defines the term as a feeling of completeness arising from one or more shared, presumably pleasant and enjoyable, past life experiences. There may be a strong emotional reaction at the first encounter with the other incarnated soul, or perhaps a feeling of having known someone before the first physical encounter, a sense of connection or even déjà vu. These feelings are at an intuitive level because they emerge into the physical consciousness from the subconscious or soul mind. The woodworking image of mortise and tenon (440-20 and 1556-2) and the idea of a complementary relationship evoke thoughts of a tight, close-knit relationship, perhaps an emotional bond with the feeling of comfort and completeness.

> *... from the mind, or the individuality of a soul, expressions are given; and thus we understand soulmate, soul-love, soul-experience. For these find companionship one with another. [Edgar Cayce reading 5254-1]*

A soul mate is not necessarily the person to whom you are physically attracted. It is more properly understood as part of the mental and spiritual relationship between individuals. Physical attraction to the opposite sex may the driving force of some individuals in the pursuit of a relationship, but that may not be a true indicator that identifies a soul mate (2988-2). Physical attraction may cause individuals to believe that their soul mate has been found, but the emotional feeling between soul mates should arise from a deeper mental and spiritual reservoir, not be confined to physical appearance or sexual desire. The readings warn that physical attraction and feelings based on outward appearances are unreliable indicators and have a potential downside as individuals discover they have quite different ideas and ideals.

There is an even greater understanding to be had in the term "soul mate" than a comfortable and close relationship between incarnated souls based on memories of shared physical experiences. Reading 2988-2 warns the inquirer to look to God as the real soul mate, not toward another person through sexual attraction or shared experiences. The highest spiritual ideal is expressed in the concept of soul mate as the relationship between a soul and its creator. The conscious awareness of God and certainty of his abiding presence within us far surpasses any sense of shared experiences or feeling of attachment or love between two persons. The sense of companionship experienced when that companion is God is beyond anything we share with any other soul. The sense of completeness that comes with soul attraction based on shared experiences during previous incarnations may be so strong as to distract the soul from its higher spiritual purpose to restore its relationship with God. It may cause the soul's spiritual progress to stall out as it focuses on shared activity with its soul mate to the detriment of spiritual activity that will draw it closer to the its creator, the ultimate mate for every soul (Todeschi, 2002).

> *But know, the soul is rather the soul-mate of the universal consciousness than of an individual entity. [Edgar Cayce reading 2988-2]*

The Death of the Soul

> *Must each soul continue to be reincarnated in the earth until it reaches perfection, or are some souls lost? (A) Can God lose itself, if God be God - or is it submerged, or is it as has been given, carried into the universal soul or consciousness? The SOUL is not lost; the INDIVIDUALITY of the soul that separates itself is lost. The reincarnation or the opportunities are continuous until the soul has of itself become an ENTITY in its whole or has submerged itself. [Edgar Cayce reading 826-8]*

It is true that souls are eternal but only within the greater context of their creation and continued existence as a portion of God. Souls are not separate spiritual organisms made by God but are a portion of God, not the whole of God, but possessing within themselves in potential all of the attributes and qualities that make God an eternal loving being. To maintain its separate existence as an individual unit of spirit with free will, the soul must submerge its self-centeredness and desire for self-gratification, its propensity to act selfishly and thereby create disharmony in the body of God. One set of choices and actions leads to companionship with God, and the other leads to separation from God, the source of life for the soul. The will of the soul can activate a desire that arises in the mind and bring it to fruition and expression in the material world, and it can act to suppress or reject a desire and stop it from becoming manifested and habituated. The will of the soul is able to change a godly mind into an ungodly mind or an ungodly mind into a godly mind. When a soul's free will is used to express and sustain selfish desires instead of godly desires, the consciousness, or mental state of the soul, is altered, and the soul loses the ability to remember or comprehend its oneness with God.

In the extreme limit of separation from God consciousness, souls, eternal in their original creation, can cease to exist. They die, or more accurately, lose their existence as independent units of spirit as they mentally and willingly distance themselves from the source of life. The loss of God consciousness by souls began in the spiritual realm, when many souls acted on their selfish desires to experience their newly granted ability to create in ways that drew them away from God consciousness. The original God-given pattern of mind in these souls was altered in ways that eventually began to impede and reduce their awareness of their creator and source of life. The original state of oneness in purpose and desire that had existed between the soul and God was disrupted and damaged. The failure of a soul to recognize and reverse its separation in consciousness from God is tantamount to a death sentence for the soul, with death meaning the loss of individual identity (will and mind) and the absorption of its spirit essence back into the greater whole, the Spirit that is God.

> *If anyone sees his brother commit a sin that does not lead to death, he should pray and God will give him life. I refer to those whose sin does not lead to death. There is a sin that leads to death. I am not saying that he should pray about that. (1 John 5:16 NIV)*

The above biblical passage indicates that most sins do not lead to death because they can be overcome through helpful prayer and presumably repentance for the sin and determination to stop repeating it. However, there is one sin that leads to death, apparently even if the perpetrator asks for forgiveness. This is a confusing piece of scripture and John does not clarify which sin falls into this category of one. John does not specify that he is speaking of death of the body as happened to Ananias and Sapphira (Acts 5:1–10), but this seems to be the common interpretation of the verse. Some commentaries suggest that when God feels a person has reached a willful, continuous, and unrepentant state, he may decide to take the person's life, and some describe God as purifying his church by removing these sinners. Both interpretations are slightly more polite ways of saying God decides to kill certain offenders. The readings would suggest a different way to interpret the verse by placing the focus on the soul, not the body. The repeated failure of a soul to forgo its selfish rebellion against its creator widens the conscious separation between soul and God. When a person refuses to accept repeated opportunities to open their mind and heart to the Christ within, that is a recipe for many cycles of physical death as they struggle with the consequences of their repeated poor choices, but more importantly it can eventually lead to perpetual spiritual separation from God and soul death, complete loss of soul consciousness and independent existence from God.

Although it is God's will that no soul should perish (2 Peter 3:9), he does not force his will upon any soul. Just as he did not force souls to remain with him in the original loving parent-child relationship, he also does not force any mature soul to leave its state of self-imposed separation to return to him. It would be impossible in any case because it could not be done without disrespecting the free will he has already granted to souls. The subconscious or soul mind, just like the physical mind, would resent such an intrusion, and the resulting relationship would never be a state of true sharing companionship. God patiently waits for us to come to our (mental) senses just as we must patiently endure the consequences of our actions if we are to effect real change in our mental and spiritual selves. As stated before, soul death refers to the loss of its individual identity and the absorption of its spirit essence back into the greater whole, the Spirit that is God. God does not desire and will not tolerate perpetual discord and disharmony within his spiritual essence. He is patient with souls, infinitely patient from the perspective of a human lifespan, but there may well be a day when he decides that enough is enough. The eternal soul may cease to exist, and it alone is responsible for its ultimate destiny.

The Purpose Of The Universe

For it is not by the will of God that any soul perishes, but with every temptation, with every trial there is prepared the way of escape. [Edgar Cayce reading 5755-2]

For the Spirit of God moved and that which is in matter came into being, for the opportunities of His associates, His companions, His sons, His daughters. These are ever spoken of as One. [Edgar Cayce reading 262-114]

From the perspective of traditional fundamental Christianity, God created the place we call earth as a life-supporting planet in the midst of a magnificent but uninviting and forbidding universe. He placed two prototype human beings, one male and one female, in a garden that would provide every need for the couple. He created the male prototype first, but then felt that the male would become lonely so made him a female prototype as a companion and helper. He promised his new biological children that they could stay in the garden if they followed the rules. Things were going well until one of them broke a rule that forbade them from learning about or understanding the difference between good and evil. The female of the new species was deceived by an evil trickster that could transform itself between a powerful nonphysical form and a snake into learning knowledge that God had declared to be off-limits, and she in turn encouraged her companion to partake of the same forbidden "fruit." When questioned by God, who already suspected the infraction because of their conscience-stricken behavior, the man blamed the woman, who blamed the snake. God was ticked off and punished them both, banishing the man and woman from the garden and setting finite limits on their life spans.

He allowed them and their descendants to make a fresh start under much more difficult circumstances where they would have to work much harder for their food, and generally made life more stressful for them to make sure they understood just how angry he was with their rule breaking. He encouraged them to procreate so as to increase the species; gave them power over all of the animals, fish, and fowl he had created to fill the earth with life; and told them to subdue the earth, often interpreted as permission to conquer, vanquish, and otherwise bludgeon the various life forms of nature into submission. Man has become very good at this, as is testified by the extinction of many species during his reign on the earth. But God let them know that all bets were off as far as their future was concerned because of their transgression of his rules. They and all of their children would have to suffer and die, to cease to exist on the earth after living a certain period of time as a human being. They were also condemned to eternal torment if they let their human frailty cause them to break some of the other rules he made up, especially the ten commandments. He did give them the promise that if their children would follow his rules and give allegiance to him, he would in turn protect them and after they died would give them new improved bodies made of a much finer nonphysical material and give them a place of honor in a great non-gender fraternity in a faraway spiritual place called heaven.

From the perspective of the readings, God created souls, a form of nonphysical life patterned in his own image, long before he created the universe or human beings. He did not and does not create souls at the moment of conception and insert them into fertilized eggs inside of women's bodies. Souls existed before the creative event that set the universe in motion, and they witnessed the magnificence of the unleashed creative force that is God. He gave the souls the freedom to think and act however they wanted even when they chose to defy their creator. The readings describe the creation of the universe as a purposeful action born out of a need to help rebellious souls come to their senses. The physical universe was created as a causal dimension where actions have consequences that must be properly understood and met by a soul so that the soul can come to the realization that selfish activity pushes it away from a loving relationship with God. When souls began to separate themselves from God by their selfish thoughts and actions, it became necessary for their survival and continuity to provide them an environment that would permit them to understand the consequences of their rebellious thoughts and actions and allow them the opportunity to see and understand their error.

> *Hence that force which rebelled in the unseen forces (or in spirit) that came into activity, was that influence which has been called Satan, the Devil, the Serpent; they are One. That of REBELLION! [Edgar Cayce reading 262-52]*

God created the universe as a time-ordered environment as perceived by the physical consciousness, the portion of the mind that communicates and interacts with the physical world through the senses of the body. This environment allows souls the opportunity to learn of their separation from him by expressing their present concept of their creator through activity and interaction with the physical world and other incarnated souls. Because the physical world of human activity is causal, the souls repeatedly come face-to-face with the consequences of their previous actions, a self-correcting mechanism that encourages them to change thought and behavior that leads to unpleasant results. In short, they were allowed to reap the consequences of the mental and physical discord they were sowing (Galatians 6:8). God set the universe in motion and gave it order using a single set of instructions, a fundamental unified law of creative force that we detect as the four physical forces of nature that govern energy and matter interactions. When the universe had evolved to the point where it had formed an appropriate planet that could support biological life, he introduced a few basic life forms activated by spirit and mind. These biological life forms were allowed to evolve, develop, and mature into the plant and animal kingdoms we see and experience in the natural world today. The history of these many diverse species is partially preserved in the fossil record of geologic strata.

> *All that was made was made to show to the sons, the souls, that God IS mindful of His children. [Edgar Cayce reading 5757-1]*

The wayward and rebellious souls began experimenting with the evolving earth environment before it was fully and properly prepared for their use. They began by creating quasi-physical thought forms in the natural physical world and by inserting themselves into existing plant and animal matter to experience the world of physical sensation and sensual excitement. This behavior and their previous rebellious behavior in spirit caused them to become trapped in the earth's environment and the surrounding and associated nonphysical realm. In response to this chaos and the ongoing need for soul rehabilitation under conditions more favorable for success, God created prototype human bodies in male and female forms, based on the body type of an existing and evolving hominid. Each human form was specifically designed to house one soul and to permit that soul to interact with the physical environment and engage mentally and emotionally with other souls encased within similar bodies. The new bodies were created as quasi-physical thought forms at five different locations scattered across the earth. They were animated by a large influx of souls who had remained true to God's original purpose and who chose to help him restore God awareness to the trapped souls. They were led by a soul named Amilius who incarnated as the biblical Adam and again much later as Jesus. As these created bodies transformed into a more solid physical form over time and genetically mixed with the existing hominid populations into which they merged, their appearance began to more closely mimic those populations. The five basic skin tones, which are still apparent in the current human population, are an environmental adaptation that was already apparent in the existing hominid populations before the entry of the new Adamic body type.

> *In that before this, we find in the beginning, when the first of the elements were given, and the forces set in motion that brought about the sphere as we find called earth plane, and when the morning stars sang together, and the whispering winds brought the news of the coming of man's indwelling, of the spirit of the Creator, became the living soul. This entity came into being with this multitude. [Edgar Cayce reading 294-8]*

The biblical story of Adam and Eve not only narrates the lives of two humans and provides an archetypical account of the influx of souls who incarnated to rescue the trapped souls but also tells the story of every soul's loss of God consciousness. Souls made intentional decisions to rebel against God just like the earthly pair who chose to eat of the forbidden fruit. The thoughts and actions of souls cause them to put at risk their position as God's favored beings, just like Adam and Eve lost their special relationship with God. The earthly couple willfully acted in disobedience to the purpose and plan that God had for them and were sent out of the garden and away from God as a result. All fallen souls, all souls who are in a state of rebellion against God, have caused themselves to be "cast out of the garden" as a direct consequence of their actions and are now required to experience life in human bodies in the material plane to introduce them to causality and responsibility for their defiance of God. Like Adam and Eve, they must become worthy of a restored relationship with God by subduing the earth, learning to subdue the transient and impermanent attractions and pleasures of a material world, and exchanging those dubious pleasures for a harmonious eternal existence as a companion of God.

All of the trapped and lost souls would eventually be required to enter the earth environment only through the biological descendants of the original prototype Adamic bodies. Every earthly life experienced by the incarnated souls either brought them closer to God or pushed them farther away from God depending on the moral and spiritual quality of the choices they consciously made. Every choice that has been made to selfishly gratify the senses of the body or to achieve power over another human has caused the soul that made the incorrect choice to face a similar situation in a future life scenario. This process will continue until the soul learns to make choices consistent with the guiding God-consciousness that is always available and accessible to the soul. Every action that conforms to the higher divine qualities inherent in a soul brings it consciously closer to God and helps free it from entrapment in the physical realm.

There have been many advances in cosmological science during the past century that have allowed scientists to follow the evolution of the universe from its early state as a miniscule high-temperature dense region of concentrated energy to a fundamental-particle soup to a vast universe of stars and galaxies. This new knowledge of the origin of the cosmos has largely been driven by scientific advances in semiconductor and other technologies that have enabled the construction of ever more accurate and sensitive methods of detecting and measuring energy from events occurring in far distant reaches of space and time. The spacetime characteristics of the universe allow every incarnated soul that is willing to approach life's experiences with patience and an attitude of faith that God is providing these experiences as opportunities for spiritual growth to recognize its separation from God. The physical universe was designed and constructed by God to provide an environment in which the Spirit of God can bear witness to our souls that he truly exists and is our creator. Paul realized the spiritual meaning of the universe and alluded to the purpose of the material world, but his words also suggest a form of predestination that isn't compatible with the readings. Our own choices are the primary factors that determine where and when we will again appear on the earth.

> *From one man he made every nation of men, that they should inhabit the whole earth; and he determined the times set for them and the exact places where they should live. God did this so that men would seek him and perhaps reach out for*

him and find him, though he is not far from each one of us.
(Acts 17:26-27)

The response of a soul to its life experiences reveals its current understanding and concept of God and its perceived relationship with God. We, as incarnated souls, have the opportunity to bear witness to the truth of God's existence and his love for us by the thoughts we entertain about him and about others, by the words we use to speak to others, and by our actions toward others and in response to the actions of others. Time and space are consciously perceived properties of the universe that place our soul's awareness and activity in a causal setting, forcing us to experience the consequences of every manifested action, not as a form of punishment but as an impartial spiritual law. These properties of the universe are also a material physical manifestation of the Spirit of God in a finite four-dimensional spacetime continuum. We can use space and time in conjunction with the spiritual attribute of patience, a spiritual quality innate within each soul, to understand God and our relationship with him. With every activity of our body and mind on the earth, we can gain a greater awareness of God and understand our true identity as souls, his spiritual children.

Time, space, and patience, then, are those channels through which man as a finite mind may become aware of the infinite.
[Edgar Cayce reading 3161-1]

The eternal soul is too precious to send into the earth for a seventy- to eighty-year crash course in the fundamentals of God consciousness only to die if it fails to quickly grasp the lessons. The traditional Christian perspective of life, death, and salvation is akin to giving a child an instruction manual on swimming, throwing the child into the pool, and telling it to learn to swim or die trying. A few will succeed and survive, but most will not. The law of God, the law that is God, is designed to offer the soul every opportunity it could possibly need as often as necessary to realize it has lost the greatest of all opportunities, the opportunity to be in a state of full attunement and companionship with its creator and Father. The earth is part of a larger institution of higher learning for the soul. Not just any kind of learning facility but one of those work-study institutions where classroom tutoring is interspersed with hands-on experience. This gives the student the opportunity to apply the bookish knowledge it has been studying to real-life situations. The earth is a place where a soul can put its knowledge into practice in an environment that lets it contribute to the welfare of those fellow souls that need help even more than itself in a community setting. It is necessary for a soul to apply spiritual knowledge in practical situations so that it becomes reinforced in the mind of the soul and part of the intrinsic nature of the soul. The knowledge it has gained is made useful and practical as it is used to help other souls that are in need. This method of learning allows pure knowledge to become applied knowledge, to become ingrained in the soul mind so that it becomes second nature and part of its normal approach to life situations. Those who absorb the knowledge offered and never use it for the good of others are described in a verse from the Book of Matthew.

> *Not every one that saith unto me, Lord, Lord, shall enter into the kingdom of heaven; but he that doeth the will of my Father which is in heaven. (Matthew 7:21)*

Souls in the spiritual realm disrupted the harmony of the realm by engaging in thought, desire, and activity that was counter to the purpose for which they were created. They willed themselves away from God by seeking selfishness (evil) and abandoning the goodness of God. Because causality is not an intrinsic property of the spiritual realm, the souls did not perceive the damage that the inappropriate use of free will was causing to their relationship with God and with the other souls that were created by God. In their selfishness and self-centeredness, the souls failed to understand that every one of them is a portion of the one, all-pervading God. Their thoughts and the subsequent activity of their soul bodies slowly deteriorated the original intimate conscious awareness and attunement they had with God until many of them were deeply absorbed and enamored with the creative power of the mind, which could be directed into selfish channels of activity without regard to the injury and defilement of the soul and God.

The souls used their God-given minds to create an alternate reality. They directed their minds to build patterns of thought and behavior that congealed into self-serving activity and lost their initial identity as companions of God with the ability to share in his creative adventures. The only way out of the mess was for the souls to experience an environment in which they could perceive the negative effect their thought patterns were having on the harmony of the celestial sphere, the spiritual realm. They needed an environment in which their thoughts would manifest and solidify, where the manifested thoughts of many souls would comingle and create an edifice of harmony and cooperation or would in their shared environment clash and cause turmoil as each soul manifested its own selfish ideals without regard to the effect imparted to other souls.

> *Remember that thy Pattern through suffering became the Savior. Through suffering, individuals meet themselves, and fit themselves for the Divine expression in their soul selves. The bodies that cause disturbance by the disobeying of the laws and activities in a material world are but troublesome things to the soul at times. LIFE IS CONTINUOUS! THE SOUL MOVES ON, gaining by each experience that necessary for its comprehending of its kinship and relationship to Divine. [Edgar Cayce reading 1004-2]*

> *In spirit the body suffers not, but there is the consciousness of all that has been before. [Edgar Cayce reading 583-8]*

The chosen method for making souls aware of their loss of God consciousness was suffering. In the spiritual realm, the soul does not endure suffering, is not aware of the suffering it causes, and does not perceive the disharmony and harm it creates when it defies God and chooses to put self before the Almighty. But the selfish activities of souls that are incarnated in the physical realm cause visible and experiential repercussions to the body and those people with whom the body has a physical, mental, and emotional relationship. Even after the body dies, the soul remembers all of the destructive activity and the suffering it caused and all of the loving and kindhearted activity and the joy and hope it brought while it was expressing itself in the earth's environment. These experiences allow each soul to understand its failure to remain true to God's plan for companionship with the created souls and to realize that it has drifted away from God in consciousness. The creative force that unleashed the formation and evolution of the physical universe was the first step on the long road by human standards to building a suitable environment where causality was the basis of the natural system. Every manifested activity by every soul that enters the earth's environment produces a ripple effect that is experienced by other souls within its sphere of influence. Soul development is both individual and communal. Each physical life offers the soul the opportunity to make changes that will bring it into better mental and spiritual alignment with its creator and into a more harmonious relationship with God. There is a spiritual purpose to the physical universe.

God Moved - The Creation Of The Universe

God moved, the spirit came into activity. In the moving it brought light, and then chaos. In this light came creation of that which in the earth came to be matter; in the spheres about the earth, space and time; and in patience it has evolved through those activities until there are the heavens and all the constellations, the stars, the universe as it is known ... .
[Edgar Cayce reading 3508-1]

Spirit, or God, is the First Cause, the essence of creative power that brought forth the physical manifestation of energy that was the origin of the universe. There is nothing that is not God, nothing that is not a manifestation of the Spirit of God. God does not create stuff, spiritual or material, that is outside of himself; he is the totality of all that exists and all that has been created. He didn't walk into his workshop like some supernatural Santa Claus and grab a little of this and a lot of that and start working on his latest invention, a new toy, a universe of energy and matter. The universe and the earth it contains is not made of some material that exists outside of the realm of God, is not some form of building material that God found lying around in his workshop and decided to use to build a new universe. The universe isn't some alternate plane of existence separate from and independent of a higher plane in which God resides.

Thus we find the manifestations of life, the manifestations of energy, the manifestations of power that MOVES in material,

are the representation, the manifestation of the Infinite God.
[Edgar Cayce reading 1158-14]

The universe exists as a form of condensed Spirit within and part of the greater realm that is the Spirit of God. This universe is a part of God, and that includes our physical bodies and all other life forms. We would never think of the lumber or metal we use to construct our houses or manufacture our machines as part ourselves, although there is a certain truth in that idea. The structure of the universe that we interact with today, that we feel as gravity and see as electromagnetic waves impinging on light receptors that coat the inner wall of the eye, and that we feel as atomic vibrations against nerve endings in the skin began as the expansion of an enormous quantity of energy and its subsequent condensation into fundamental particles, atomic nuclei, and finally the atoms that form the basic units of the material we call matter.

The universe did not create itself. It is not the result of a spontaneous eruption of matter and energy out of emptiness. The universe was created by God but not on a whim or without purpose. The universe was not created by God in some distant past for the enjoyment and pleasure of generations of human beings yet to come. The universe is part of a plan that God put into place after souls, using their free will in selfish and inappropriate manners, began to lose conscious awareness of their creator and their original intimate relationship with him. The universe was designed specifically as a causal learning environment in which souls could experience the consequences of their actions, recognize that ungodly actions separate them from their Creator, and begin the long process of restoring that original relationship that existed "in the beginning," long before the universe was created. The lifespan of the universe is huge but still finite. Because it is finite, it must have had an origin and will have an ending. The origin of the universe has its roots in the activity of the Mind of God, a Creative Force that initiated the expansion of a spacetime dimensionality within the spiritual realm designed to contain the environment necessary to allow and constrain physical evolutionary processes that would ultimately led to the formation of the various elements, molecules, and compounds we experience as matter.

> *... for all force, all power, EMANATES from one source. So, as has been given, that which makes up even the electronic energy - that man knows as electrons or energy in electricity, is of God itself. [Edgar Cayce reading 4757-1]*

The universe was created by God but not in the conventional sense that we as humans would imagine, expect, or anticipate. As human beings living in this universe, we are immersed in physical matter and are used to interacting with the physical world through our senses. We also vicariously experience matter and energy through the mechanical and electronic instruments we have devised to explore and harness our environment. This ongoing experience causes us to unconsciously explain the unknown in terms that fit into our known circumstances, experiences, and beliefs, a process called anthropomorphism. Did God really go out to his workshop and get a spiritual hammer and a box of divine nails and with his blueprints in hand started hammering out a universe? Of course not, but we do love our anthropomorphisms. We tend to interpret impersonal events that are not of human origin in terms of human or personal characteristics with which we are familiar and understand. The universe was not created by God using material we are familiar with by moving himself outside of his spiritual realm into a region of matter external to himself, as we would imagine a carpenter arriving at a building site and picking up a hammer and nails to construct a house. There are no raw materials outside of God that he can use whenever he dreams up a new construction project to satisfy his creative drive. There is no such thing as outside of God. The universe was created using the only building material that exists. It was created from and through Spirit, for Spirit is the First Cause of all that is created, the original substance, not created from any other substance, and the only substance from which all else is derived. God is Spirit. God created the universe out of himself. The Apostle Paul glimpsed this reality, and his thoughts are expressed in the Acts of the Apostles.

Paul may not have had the same concept of spacetime as envisioned by modern physicists, but he did understand a fundamental principle of the universe in relation to spirit.

> *"For in him we live and move and have our being." As some of your own poets have said, "We are his offspring." (Acts 17:28)*

God is infinite, eternal, and encompasses all that is and was and will be. But because the universe is a created thing that is not constantly being manipulated and controlled by the mind, it is finite and transient, having a beginning and an ending. Two ideas are presented in this Bible verse that describes our life and origin. The first idea, of living and moving in God, coincides precisely with the readings; we do not just exist within a physical universe created by God, but we exist within the totality that is God. The second idea that we are his offspring, in light of the first idea, suggests that we are more than physical bodies and alludes to our creation as spiritual children of God, as souls. Our souls are actually composed of spirit, the essence of God, and we exist in God as part of God, with the full potential to convey the attributes and characteristics of God into our relationship with others.

The beginning of the creation process that eventually led to our present-day universe is expressed in the readings in a simple yet profound way as "God moved, the spirit came into activity" [35081]. What did the source of the readings mean by God moved, or as stated in other passages, the Mind of God moved, and the universe came into being? We are used to thinking of movement or motion as describing a change of position in space over a specified time interval; that is, motion requires and takes place within the framework of space and time. But this activity of God preceded and caused the existence of space as we understand it and time as we perceive it by our physical consciousness. Notice that this same expression is used in the readings to describe the manner in which God created souls (263-13). The soul of Jesus, having remained loyal to God's will and being the firstborn of all souls, was a co-creator with God in the spiritual and mental activity that set into motion the events necessary to condense spirit into concentrated energy and establish a force law that would cause an expansion of the dimensions science now understands as spacetime. The soul of Jesus was a co-creator of the universe and was the leader of the group of souls that later attempted to restore the knowledge of God to rebellious souls that became entangled in the world of matter and energy.

> *God, who at sundry times and in divers manners spake in time past unto the fathers by the prophets, Hath in these last days spoken unto us by his Son, whom he hath appointed heir of all things, by whom also he made the worlds. (Hebrews 1:1-2)*

So then, what does it mean for God to move in the sense of initiating the creative act that ultimately culminated in the universe we know and the earth we inhabit? Does the statement more accurately describe a mental decision, or a resolve to bring a choice into activity, rather than a description of spiritual locomotion or the movement of a spiritual being? Creation, according to the readings, is an act of will that directs the mind to utilize and crystallize spirit around a specific thought. Spirit moves, is set into creative motion, by the will activating a desire held in the mind. Everything in the universe evolved from Spirit in motion, Spirit active within the confines of a unified space and time dimensionality which expresses itself in materiality as energy, and which we experience and detect as electromagnetic radiation. The Creative Force that drove the initial expansion of spacetime and was a part of that spacetime expansion was in some way the Spirit of God, the same substance that is the essence of God, and therefore the universe itself is both a part of God and a creative expression of God. In this creative act God conceived and set into motion the universe wherein the soul would be limited in the scope of its abilities but could become more consciously aware of the impact of its mental choices on others by residing in a causal environment. The consequences of every creative act of the soul would be reflected back to that soul. This environment would enable the soul to meet itself, to come face-to-face with its own decisions and shortcomings and, in doing so, become aware of its separation from God. He perceived a physical universe as the best environment in which souls would have the opportunity to eventually come to realize their separation from him, to learn to correct and remove selfish patterns of thought and unite with him in the state of true companionship he desires of them.

The preparation of this environment for soul development did not occur in an instant, or even in the seven days biblical literalists promote. Cosmologists have made great progress in the past fifty years in understanding the evolution of the universe from the original rapid expansion of its energetic core to the current distribution of stars and galaxies. The menagerie of fundamental particles that appeared with improvements in particle accelerator designs and ability to generate higher-energy particle-particle collisions has finally resolved into a more manageable four basic particles that are the atomic-matter building blocks of today's universe and that form all physical objects. The creation of matter has been underway for at least 13.8 billion years as we measure time on the earth and will continue for billions of years to come. We know almost nothing about the mechanism by and through which God initiated the creation and expansion of the universe we inhabit, but we know much about the physical processes that controlled the earliest moments in the evolution of the universe.

> *Just so have we seen and comprehended how that there is the Father, the Son, the Holy Spirit. The Spirit is the movement; as when God the First Cause - called into being LIGHT as a manifestation of the influences that would, through their movement (light movement) upon forces yet unseen, bring into being what we know as the universe - or matter; in all its forms, phases, manifestations. [Edgar Cayce reading 262-119]*

It would be exciting and informative to view this creative act from outside the physical universe looking in, which was our soul's original perspective, but we are currently unable to take that perspective because our souls are now confined to a physical body residing within the created universe without direct access to the subconscious mind. From this side of the creative act we can only use the senses of the physical body to observe remnant features of the universe that still hold evidence of the original creative act or perhaps use the sensitive electronic instruments we have learned to build that are able to extend the range of the energy spectrum that we can observe far beyond the energy range that our senses can directly detect. We probably wouldn't be able to understand how God did it even if he tried to explain it, but we can try to gain some understanding by extrapolating our observations of the cosmos and knowledge of nuclear and atomic physics to as close to the start of the universe as is possible. Today, physicists have a good understanding of the physical processes and laws that control the formation and interaction of the stars, planets, galaxies, black holes, and other strange objects that have been discovered and observed scattered across the vastness of space. We have the capacity to make precise astronomical observations, to collect a vast amount of information on the behavior of atomic matter and energy, and to analyze these data to form an intellectual understanding of the origin and evolution of the universe using logic, mathematics, and tested physical laws.

Energy and Particles

Early theories of the origin of the universe suffered from a serious problem. Scientists had concluded that the energy equivalent of all the matter and energy that makes up today's entire universe must have existed at every stage of the early universe. However, this would mean that the energy density and temperature of the universe increases exponentially the closer one gets to the original event that initiated the expansion of the universe. This is not physically reasonable. Most people today have probably heard of the Big Bang that is supposed to have initiated of the creation of the universe. But the name is a misnomer that suggests an action that has nothing to do with actual events at the earliest moments in the formation of the universe. Even scientists often describe the initial creation event as the Big Bang because of historical reasons and early attempts to understand the process, but that term is inappropriate.

The term "Big Bang" gives the impression that something akin to a super massive hydrogen bomb detonation was the defining event of the beginning of the universe. It suggests an explosion, an enormous and violent outward burst of matter and energy from a dense central core in which all that exists in the universe was concentrated. In reality there was no bang or explosion, there literally was nothing for an explosive force to expand into because there was no waiting volume of space to receive a burst of matter and energy. There could be no ejection of matter in any case because neither subatomic matter nor atomic matter existed in the early universe. A more realistic scenario has the universe beginning as an unfathomably small region of nascent spacetime continuum that experienced a phenomenally rapid expansion before settling into the universe we know today. Science is not yet able to precisely describe the conditions that preceded or caused this rapid expansion of energy and has no knowledge of the substance that existed as predecessor and source of this concentrated energy. The readings equate this prototype energy with the movement of Spirit into physical dimensionality.

This initial seed of energy precipitated an abrupt, extremely brief, and nearly exponential expansion of the universe during the earliest stages of its development. The universe continued to expand, but at a much slower pace, following the rapid expansion phase. The energy density and temperature of the universe as a whole quickly decreased during this earliest phase of rapid expansion, eventually producing the conditions that ultimately permitted the condensation of quarks and electrons and the formation of atoms. What might this initial condition of extreme density and temperature tell us about the God who set these conditions in motion? These conditions are suggestive of certain passages in the Bible that describe disincarnate entities such as when Elijah and Moses conferred with Jesus during the transfiguration on the mount (Matthew 17:1–3 and Mark 9:2–4). They were all brilliantly white with an intensity that made them difficult to observe with the physical eyes. This may be an indication of a high concentration of radiant energy, a brilliant white light so bright as to nearly blind any individual observing the phenomena.

The current best description of the fundamental particles that make up the universe is called the Standard Model of Particle Physics. It includes four matter particles (they have mass) called fermions, each of which occurs in three different energy states, or families, for a total of twelve matter particles. Only the four particles of the lowest energy state can exist at the temperatures, pressures, and densities found under natural conditions in the universe today. The eight higher-energy versions of these particles existed in the earliest environment of the universe when temperatures, pressures, and densities were much higher, and can briefly exist in particle accelerators where they can be created by matter-matter collisions. Each of the twelve matter particles has an antimatter twin. Antimatter particles are not the exotic building blocks of some alien universe but in the simplest terms are the charge opposites of each particle. For example, the antiparticle of a negatively charged electron is a positron, a particle having the same mass as an electron but a positive charge. The eight higher-energy antimatter particles existed in the very earliest moments of the universe. The antiparticles associated with the four lower-energy matter particles active in the current universe can exist briefly today during particle-pair production from electromagnetic energy but quickly annihilate with their matter twins to recreate energy. Scientists have named the four basic matter particles that are the building blocks of today's universe the up quark, down quark, electron, and electron neutrino.

The exchange of energy between particles with mass are regulated by massive and massless force carrier particles called bosons. From the weakest to the strongest they are: the graviton that mediates the gravitational force of attraction between all matter particles; the W and Z bosons that carry the weak nuclear force and are responsible for radioactive decay; the photon that carries the force of electromagnetic attraction or repulsion between charged particles; and the gluon that mediates the strong nuclear force (also called color force) that binds quarks into protons and neutrons. The most familiar force in our everyday lives, gravity, is not part of the Standard Model of Particle Physics because scientists have been unable to mathematically incorporate the current best description of gravity into this theoretical framework for particles. And then there is the Higgs boson, the latest particle to be included in the Standard Model. The Higgs field and associated Higgs boson are postulated to have provided the mechanism whereby massless particles in the early universe acquired mass and evolved into the fundamental matter and force particles.

The four force carriers are associated with the four fundamental forces and force fields that are familiar to us today: the gravitational force, the weak nuclear force, the electromagnetic force, and the strong nuclear force. Each force field has a characteristic range of effective action and strength. Gravity is the weakest force by far, being only about 10^{-38} as strong as the strong force (that is, weaker by a factor of one hundred trillion trillion trillion), but it has an infinite range. Readers who are unfamiliar with the scientific notation method of writing numbers may want to refer to the brief Scientific Notation section at the end of the book before continuing. Gravity is the attractive force that draws together thinly distributed atomic matter in space into matter groups such as nebulae, galaxies, and stars. The weak force is a short-range force, primarily moderating radioactive decay, and its strength is about one ten thousandth that of the strong force. The electromagnetic force is about one hundredth the strength of the strong force, and it also has an infinite range. It is both a repulsive force and an attractive force and is instrumental in binding negatively charged electrons to positively charged atomic nuclei to create atoms. The strong force is the strongest of the four forces, being very powerful but limited to only a very short effective range. On the nuclear scale, its strength is repulsive over extremely short distances but becomes strongly attractive over moderate distances and rapidly loses strength with distance greater than a single nuclear radius. It is associated with the binding of quarks into protons and neutrons and the clumping of protons and neutrons into atomic nuclei.

What have we learned from this side, the "inside" of the universe, about its construction and evolution? God unleashed his new universe as an expanding foundational framework of spacetime which manifested as an exceedingly hot and dense core of energy in the form of electromagnetic radiation and eventually atomic matter about 13.8 Bya. The temperatures and pressures in this proto-universe are unimaginable compared to our usual environment and experience. Temperatures were as high as 10^{30} Kelvin (32 Fahrenheit = 273.15 Kelvin) and the energy density was at least 10^{80} tons per cubic centimeter. This activity at the beginning of the physical universe is well described in the readings.

God moved, the spirit came into activity. In the moving it brought light, and then chaos. [Edgar Cayce reading 3508-1]

Know then that the force in nature that is called electrical or electricity is that same force ye worship as Creative or God in action! [Edgar Cayce reading 1299-1]

God, spirit, and electromagnetic energy are intimately linked. The readings imply that electromagnetic energy is a form of Spirit in motion. Exactly how spirit moves is left to the reader's imagination. It will be interesting to see if science can eventually look deeper into the nature of electromagnetic energy and find the underlying reason that causes it to always move through the empty regions of the universe, those regions not occupied by matter, at a constant enormous speed, the speed of light. Electromagnetic energy can never rest, never stop moving, and never slow down. It may appear to slow down when it travels through atomic matter such as glass, but only because it is constantly and repeatedly being absorbed and re-emitted by atoms.

The basis, then: "Know, O Israel, (Know, O People) the Lord Thy God is One!" From this premise we would reason, that: In the manifestation of all power, force, motion, vibration, that which impels, that which detracts, is in its essence of one force, one source, in its elemental form. [Edgar Cayce reading 262-52]

Reading 262-52 also emphasizes that the First Cause as Spirit set the universe in motion and that Spirit is expressed in the physical realm as electromagnetic energy. It discusses the confusion that arises in the minds of men as they learn to harness the force of electromagnetism to supply electrical power for the benefit and convenience of our lives. Because we can influence and harness electromagnetic energy and can describe it mathematically, we think of it as a purely physical phenomenon and do not recognize the spiritual basis of this substance. When the mind of man directs spirit as electrical energy into manifested form, he believes that he has power and influence over this force, but in reality the knowledge that allows us to control electromagnetic energy and use it to manipulate matter and energy reactions in the material world is only a poor understanding of the true nature of the force. Our mastery over the physical phenomena associated with electricity blinds us to the possibility that there is more to it than can been seen or felt through the senses and gives us a feeling of power over nature that relegates it to something less than us, something outside of us, and something to use but not truly comprehend. Electrical energy is less than the soul in its physical manifestations but as great as the soul in its spiritual origin.

Electricity or vibration is that same energy, same power, ye call God. Not that God is an electric light or an electric machine, but that vibration that is creative is of that same energy as life itself. [Edgar Cayce reading 2828-4]

Phase 1: The scientific understanding of the nature of the physical universe from the first moment of its formation until about 10^{-43} seconds after its birth has developed slowly over time. For convenience this period in the history of the universe is called Phase 1, the first of five phases described herein, but note that the scientific literature does not use this terminology. During Phase 1, time (duration or time interval) did not exist in the sense that we understand it, and space (distance or spatial interval) had no dimensional meaning in the sense that we understand it. The concept of time that we use to measure duration and motion in the universe has no meaning in physics before about 10^{-43} seconds. The fundamental unit of time, the Planck Time, is set by the laws of physics at 5.4×10^{-44} seconds. It doesn't make sense in physics to speak of any smaller time interval. Similarly, the fundamental unit of space that is used to measure distance and motion, the Planck Length, is set at the value of 1.6×10^{-35} meters. It doesn't make sense in physics to speak of any smaller spatial interval. The basic physical laws of the universe do not have meaning at temporal durations and spatial intervals less than the Planck Time and Planck Length.

A single force ruled this early proto-universe, a universal force that was the unification of the four physical forces we now observe. The readings state that, "all force is vibration, as all comes from one central vibration" (900-422), suggesting that this primordial universal force is an aspect or part of the one Creative Force that the readings use as a synonym for God himself. It also sets the stage for a persistent theme of the readings that indicates all spiritual and physical forms have their own fundamental vibration. This universal force may be God's hand on the wheel, his method of guiding the formation of inanimate matter and the evolution of the features of the universe to the form we observe today without the need to consciously micromanage every aspect of physical creation.

> *... the universal [force] being that element through which all becomes manifest in a material world, or a spiritual world. As would be illustrated in the prism separating the elements of light, and showing the active principle of given light or heat in its action, by deflection from the given law, the universal forces are as such. [Edgar Cayce reading 900-17]*

A prism, a transparent material (usually glass) through which light can pass, separates white light into its component frequencies. As a light wave enters a prism at an angle less than ninety degrees to its surface, one part of the leading front of the wave can strike the prism earlier than other parts. This causes different photons of the wave front to interact with electrons in the prism at different times and effectively changes the direction of propagation of the wave front as it passes from, for example, air to glass. The angle of the change depends upon the refractive characteristics (index of refraction) of the two materials traversed by the light wave. The refractive index is a function of light frequency, thus white light (a composite of all visible light frequencies) or any light composed of different frequencies (colors) is split into its various components. The readings liken the spirit-material world interface at the instant of creation of the universe to a light beam striking an air-glass interface. The single universal force of the spirit world that is the active force in the creation of the universe became split into individual component forces that we identify and label as the four known physical forces.

Expansion and Division of Forces

Phase 2: Our knowledge of the early universe from about 10^{-43} seconds after its onset to about 10^{-14} seconds after its onset is limited and speculative, but is constrained by extrapolation of known laws of energy and particle interaction. The time interval of this period of expansion seems too ridiculously small to worry about, but to think that nothing of importance can happen in an interval of time that short would be incorrect. The interval encompasses a range of time that spans a factor of 10^{29} (equivalent to a time interval from one second to three trillion billion years). We do know that this phase was characterized by a brief period of rapid expansion of the spacetime continuum from about 10^{-36} seconds to about 10^{-33} seconds after the birth of the universe. Before this expansion the universe was held in potential in the form of an inflation field, also referred to as the grand unified Higgs field. The values of the inflation field were not yet determined but existed as a quantum froth of potential values, sometimes called a false vacuum state because the average potential energy of the field was not zero. As this quasi-stable initial condition settled into the stable field values that define the average zero-energy vacuum of the spacetime continuum in our universe, the false vacuum energy was released into the forming spacetime structure and then into the creation of radiation and matter. Inflation caused the size of the universe to increase by a factor of 10^{26} (one hundred trillion trillion). During the expansion, temperatures in the universe plummeted by twenty-three powers of ten and the density of the universe fell by fifty-five powers of ten. These enormous changes altered the behavior of energy in ways that are virtually unimaginable to the mind, and present-day science is incapable of accurately predicting the behavior of the universe during these changes.

Cosmologists speculate that this inflationary burst that began the formation of the universe was an event that occurred within a preexisting universe rather than the creation of a new universe from nothingness. The readings are consistent with this idea, indicating that the preexisting universe was a form of Spirit, the substance that is God. The force driving the expansion and imparting motion to the energy contained within the spacetime continuum is the physical manifestation of Creative Force that we call gravitation. Matter and energy generate gravitational attraction but the release of the false vacuum energy generated a pressure that caused gravitational repulsion. Gravitational repulsion dominated the universe during the brief moments in which the inflation field collapsed and the immense outward force of the pressure caused the spacetime continuum to rapidly inflate. By the time the field had fully released its energy, the universe was primed for the formation of the first matter particles.

The term "Creative Force", a frequently used name for and description of God in the readings, describes God as the impetus behind all manifested forms. It also seems to be an appropriate description of the original unified force that drove the initial expansion of the universe and evolved into the four basic forces of nature (281-3). The single force that controlled events at the onset of the Phase 1 universe changed in both magnitude and character as the spacetime continuum expanded and temperature and density rapidly declined during the Phase 2 universe. By the onset of Phase 2 at 10^{-43} seconds, the force of gravity was already an independent force associated with the curvature of spacetime but the other three forces were still unified. Physicists have a good understanding of the force carriers that mediate particle interactions and have developed a theory called the Grand Unification Theory to predict the behavior of the combined weak, electromagnetic, and strong forces in these early times. Because the extremely high temperatures and energies necessary to mimic the conditions that existed when these forces were united are beyond the capability of present-day particle accelerators, no experiment has verified the predictions of the Grand Unification Theory. By 10^{-35} seconds the strong force had separated from the electroweak force, the unified force composed of the electromagnetic and weak force. The properties of the electroweak force are well understood and are described by the Weinberg–Salam theory. The strong force was not very strong when it first separated from the electroweak force, but as the universe evolved and cooled it eventually tripled in strength to become the most powerful of the four forces. It is the force that binds quarks together to form protons and neutrons and binds protons and neutrons together to form atomic nuclei.

Inflation explains how the initial events within our nascent universe created conditions favorable for the formation of the fundamental matter and force particles. However, there is more than one version of inflationary theory, and some of those theories allow for repeated instances of inflation where space expands rapidly then slows, allowing energy to coalesce into particles. Each of these inflationary events can form a separate universe, and they are collectively called an inflationary multiverse. In inflation theory, quantum uncertainty means that each inflationary event will begin with a different energy and will produce a different outcome. Each instance could create a unique universe with its own characteristic physical laws and fundamental particles. Some of these universes might be quite similar to our familiar home universe but others could have physical laws and structure that would appear quite foreign when compared to our universe (Greene, 2011).

And There Was Mass

Phase 3: During the period from about 10^{-14} seconds after the birth of the universe to about 1,000 seconds after the birth of the universe, a time interval of about sixteen minutes, the universe continued to expand but not as rapidly as it did during inflation. By 10^{-14} seconds, the new universe had expanded enough to cool and lower its pressure and density to levels that are comparable with conditions simulated in modern advanced particle accelerators. Experiments conducted with these accelerators verify that the Standard Model of Particle Physics accurately describes mass and force particle interactions under these conditions. This interval encompasses a range of time that spans a factor of 10^{17} (equivalent to a time interval from one second to three billion years). This is an exciting time in which the building blocks of the current universe are created.

> *And there is no vacuum, for this, as may be indicated in the universe, is an impossibility with God. [Edgar Cayce reading 3161-1]*

The vacuum of space before any matter had formed was not dormant, empty, or a state of nothingness but was filled with the field that physicists call the Higgs field. This field existed from the first moments of the birth of the universe but the property of the field that is active during this phase of its evolution is the electroweak Higgs field, not the grand unified Higgs field. From the point of view of the readings, it makes perfect sense that there can be no true state of nothingness anywhere in the universe because the universe, like all else that exists, is a part of God, and there is nothing that exists outside of or separate from God. A region of nothingness would be a region devoid of matter, energy, and spirit and therefore God, but there is nothing that is not God and God is not absent simply because matter and energy are absent. Before about 10^{-10} seconds the vacuum energy, or energy froth, was in a constant state of transformation out of which virtual particle-antiparticle pairs were briefly created from the vacuum energy, and particle pairs were constantly annihilated to reform vacuum energy.

Virtual doesn't mean not real. It means existing only for the briefest moment in the universe before pair annihilating into vacuum energy. Quark-antiquark pairs were preferentially produced in the early stages of the universe expansion when temperature and energy density were higher and electron-antielectron pairs were preferentially produced shortly thereafter as the temperature decreased. The Higgs field was too hot and too energetic to interact with these particles. The particles produced in this most basic form of our universe were massless, traveled at the speed of light, and had extremely brief life spans.

As the universe cooled, the Higgs field reached a point where it experienced a phase transition; it condensed and acquired a nonzero average energy. This transition changed the properties of the Higgs field. All particles and force carriers of the universe move through the Higgs field. Most particles and force carriers experience inertial resistance when they accelerate through the condensed Higgs field, and each particle and force carrier reacts differently to the field. They don't feel this resistance when they are moving at constant velocity, only when they are accelerating through the field. This is precisely the definition of mass; the resistance of a body to acceleration when a force is applied to it. At about 10^{-10} seconds particles and some force carriers began acquiring mass. The electromagnetic and weak forces were unified and indistinguishable before about 10^{-10} seconds because photons of the electromagnetic force and the W and Z bosons of the weak force all were massless and had identical properties at the existing temperatures. Their properties, and therefore their effects, were indistinguishable. After about 10^{-10} seconds, the two W bosons acquired a mass about eighty times the mass of a proton and the Z boson acquired a mass about ninety-one times the mass of a proton, but the photon was still able to move through the field completely unaffected and did not acquire mass. Therefore the action of the electroweak force on particles was altered and the combined electroweak force became the separate electromagnetic and weak forces we observe today.

What is light? That from which, through which, in which may

*be found all things out of which all things come. Thus the
first of everything that may be visible, and earth, in heaven,
and space, is that light - IS that light. [Edgar Cayce reading
2533-8]*

This profound reading might have been difficult for Cayce and those first researchers of the readings to fully comprehend or reconcile with science. It was given in 1944 as part of an explanation of the attunement of physical consciousness to the Christ, a mere twelve years after the first antiparticle was discovered when scientists were beginning to develop an understanding of the concept that light, electromagnetic energy, could spontaneous form quark-antiquark and electron-antielectron pairs under certain conditions. It seems unlikely that these discoveries or their implications would have been widely known to the public or those studying the readings. The reading was received well before cosmologists understood that high-energy radiation in the form of X-rays, gamma rays, and perhaps unnamed bands of electromagnetic waves of even higher frequencies were the dominate form of energy in the earliest universe. This primordial radiation is the precursor of all the atomic matter and electromagnetic radiation in the universe today. What we call visible light is a lower-energy form of electromagnetic radiation that has a range of wavelengths from about 380 billionths of a meter to about 700 billionths of a meter and are the range of wavelengths (colors) visible to the human eye. The light waves that fall on our eyes are not the original electromagnetic radiation from this early universe but are photons produced by the interaction of electromagnetic energy with free electrons and bound electrons in atomic matter. In a sense, they are descendants of the original radiation.

As the universe continued to cool and particles acquired mass, it became more difficult for the quark-antiquark pairs to re-annihilate to form electromagnetic energy, and the quark particles begin to interact more frequently. Quarks cannot exist in isolation as free particles. Two quarks can combine briefly but only exist together as unstable heavy particles having short life spans. Three quarks can exist as a stable unit and, as the universe continued to cool to about one trillion degrees Kelvin about 10^{-5} seconds (ten microseconds) after the original creation event, quark triplets started condensing into protons, antiprotons, neutrons, and antineutrons. Shortly thereafter, the proton and neutron matter-antimatter pairs annihilated, but because of an as yet unexplained imbalance between the numbers of protons and antiprotons, about one proton was left standing for every billion protons that annihilated. The same phenomenon left behind about one neutron for every billion neutrons that annihilated. Suddenly, the new universe was filled with protons and neutrons, the basic building blocks of atomic nuclei.

The newly formed protons and neutrons were not stable. The neutrons were constantly emitting neutrinos and converting to protons, electrons, and energy. The inverse process was also occurring, protons were constantly absorbing antineutrinos and energy and converting to neutrons and antielectrons. During the time of these reactions, the number of protons and neutrons remained approximately the same, but as the temperature cooled by another factor of ten, there was less energy available for the proton to neutron conversion, and the neutrinos decoupled from the reactions and became free to propagate across the universe without being stopped by collisions with protons and neutrons. This stabilized the ratio of the neutrons to protons at the current 1:7 ratio and allowed the electrons and antielectrons to initiate the next important step in the formation of matter. By this time the universe had been around only about one second but had undergone tremendous changes. Its temperature had fallen by a factor of twenty powers of ten, its density had decreased by a factor of eighty powers of ten, and spacetime had expanded by a scale factor of twenty powers of ten.

Electrons and antielectrons were being created and destroyed in the proton-neutron conversion reactions, but after the neutrino decoupling, these reactions were much less frequent. Electron and antielectron pair production from electromagnetic energy was contemporary with and after the time of quark-antiquark production and the confinement of quarks in protons and neutrons, especially as the lower temperature and energy favored the creation of the lighter particles. By ten seconds into the evolution of the universe, the sea of electrons and antielectrons begin a mass annihilation event that left about one electron remaining for every one billion electron-antielectron pairs that annihilated to produce energy, mimicking the same process that caused the antiquarks to be removed from the universe. The reason for the tiny excess of electrons over antielectrons is not well understood. This one-in-a-billion matter-antimatter imbalance in the early universe allowed a matter-dominated universe to be formed and atomic matter as we know it to be created; otherwise, the present universe would be entirely composed of photons moving through the spacetime continuum at the speed of light. The universe now has the basic subatomic particles needed to create atoms; protons, neutrons, and electrons.

The last significant event in Phase 3 occurred during the interval when the universe was about 100 seconds to 1,000 seconds old. It was a series of nuclear reactions in which a portion of the recently formed protons and neutrons combined into stable helium-4 nuclei (a bound cluster of two protons and two neutrons). By this time in the evolution of the nascent universe, the temperatures and pressures had reduced considerably and conditions in the universe were becoming similar to those in the core of stars. Particle and energy interactions under these conditions are fairly well understood in physics, and therefore the scientific understanding of the early universe from this time onward is quite good. The nuclear reactions in the core of stars are similar to those of this period, but the temperatures and densities were still too high for the ongoing reactions to be identical to those in stars. At the beginning of this short period, the nuclear composition of the universe was about 88 percent protons and 12 percent neutrons. When the brief flurry of reactions was completed, the nuclear content of the universe was 75 percent protons (the nuclear cores of hydrogen atoms), 25 percent helium-4 nuclei, and trace amounts of neutrons, deuterium nuclei, helium-3 nuclei, and lithium-7 nuclei. The reactions that started the process were proton-neutron reactions, whereas the equivalent process in the thermonuclear fusion furnace of a star begins with proton-proton reactions because free neutrons are not present in the core of stars. The electrons that occupied space among these nuclei were not able to combine electrically with the protons to form hydrogen atoms or with the other nuclei to form light elements because they were still too energetic in this hot and dense universe to remain bound to the nuclei.

It might be difficult to believe that all of the creative activity described in the second and third phases could really have been accomplished in only about 1,000 seconds. The universe expanded by a factor of at least one hundred trillion trillion, formed quarks and electrons, created protons and neutrons, and brought protons and neutrons together to create the nuclei of the first few elements. All of this was completed in the first 1,000 seconds of a 13.8 billion-year-old universe. Is this even possible? Dr. Mark Whittle, a talented cosmologist who prepared and presented a thirty-six-lecture series for The Teaching Company on the science of cosmology, delivered an eye-opening comparison of the relative speeds of atomic and nuclear processes. During one eye blink, molecules in the eye vibrate ten billion times; during each molecular vibration, electrons in the atoms of those molecules orbit one million times; for each electron orbit, protons in the nuclei orbit one million times; for each proton orbit, quarks in the protons orbit one million times (Whittle, 2008). Electrons, protons, and quarks do not actually orbit nuclei or each other in the conventional meaning we naturally associate with that word. The concept and analogy is imperfect but can still be a useful way to imagine the motion and speed of subatomic particles. Nuclear and atomic processes occur at rates that are unfathomable to us because they are so fast compared to anything our senses can detect or our physical mind can comprehend. An eye blink averages 100 to 150 milliseconds according to University College London researchers and between 100 to 400 milliseconds according to the Harvard Database of Useful Biological Numbers.

During an eye-blink of 100 milliseconds, quark triplets orbit within protons and neutrons about ten to the twenty-eight (10^{28}) times and electrons orbit the nuclei of atoms about ten to the sixteen (10^{16}) times.

Therefore, it takes about 10^{-29} seconds for a quark to complete one orbit, which is incredibly fast compared to the nearly 10^{-14} second time span of the Phase 2 universe and the nearly 1,000 second time span of the Phase 3 universe. A quark could have completed one thousand trillion revolutions during the entire time span of the Phase 2 universe had nuclei existed then, and it could have completed one hundred million trillion trillion revolutions during the entire Phase 3 universe. By comparison, it takes the slower electrons about 10^{-17} seconds to complete one orbit of a nucleus. An electron could have competed only about 1,000 revolutions of a nucleus during the entire time of the Phase-2 universe, but it could have competed about one hundred billion billion revolutions during the Phase-3 universe had atoms existed during those phases. The nearly 1,000-second duration of the Phase 3 universe when protons and neutrons were created was huge relative to the frantic activity of quarks and electrons. What appears to us as insufficient time for anything of significance to happen in the early universe is an enormous span of time relative to the rates at which nuclear and atomic reactions and the associated particle motion and energy transfer occur.

What would it feel like to blow into a balloon and watch it expand from the size of a marble to the size of our solar system in an immeasurably small fraction of a second? Did God feel a rush of excitement as the universe expanded at such an unfathomable rate and swelled and unfolded until particles of matter precipitated out of the energy racing through the spacetime continuum like the droplets of a fine mist condensing in the cold air of a late fall morning? Droplets of condensed energy, tiny bundles of energy wrapped up tightly with an electric charge balanced against another drop of energy with the opposite charge. How did the particles gain charge? There was apparently no charge in the packets of energy that produced the particles. What caused the one-in-a-billion imbalance in paired matter-antimatter particles that allowed a material universe to survive the cataclysmic annihilation of all antimatter particles? What does God see when he looks at the universe? Since he doesn't have physical eyes, does he perceive an entirely different spectrum of electromagnetic radiation or some form of energy we have yet to discover? How does a being see or perceive something that is created out of the essence of itself? Does he appreciate the beauty of his own creation from a perspective that we can only imagine now but that we will perhaps share and understand after we pull away from our physical body and view the universe through the soul equivalent of the physical senses?

Building Blocks of the Universe

Phase 4: This phase encompasses the period from about 1,000 seconds after the initial creative event to the formation of the first atomic matter about 400,000 years later. It might seem as if particles dominated the universe by the beginning of Phase 4 after the recent formation of free electrons, protons, helium-4 nuclei, and trace amounts of other light atomic nuclei, but in reality this was still an era of radiation dominance. There was still no atomic matter at the beginning of this stage of evolution, only a fully ionized gas or plasma. Electrons that tried to bind to protons or helium nuclei were too energetic or quickly ejected from the bound state by collisions with energetic photons from the surrounding dense sea of electromagnetic radiation, preventing the formation of stable atoms. It would take 400,000 years after the formation of free electrons and atomic nuclei before the universe would lose enough energy through expansion and cooling to allow the free electrons to become bound to the atomic nuclei and form the first atoms, the lightest elements. Even then, the density of electromagnetic radiation would be about 1.5 billion times larger than the density of the electrons and nuclear matter in the universe.

Photons could not travel long distances under these conditions because they were constantly being scattered by collisions with free and intermittently bound electrons. An observer standing in this early universe would see a uniform glow like is seen when headlight beams reflect from water droplets in a fog. The temperature of the universe at this time was about 3,000 Kelvin, which would cause the glow to appear orange or red-orange in color and its intensity would have been about a trillion times as bright as we typically experience in the mid-day sun today. But all of this was about to change in a way that would drastically alter the nature of the universe. Over a period of about 80,000 years, as the spacetime continuum expanded and the energy of the radiation decreased, electrons were able to condense onto nuclei and remain in the bound state without being immediately ejected by a photon. Suddenly photons could travel across the universe unimpeded except by increasingly rare collision events with the newly formed atoms. The fog lifted, and the universe was revealed to anyone who could observe it.

This burst of unconstrained radiation, called the cosmic background radiation, is still propagating through the universe and has been carefully mapped by cosmologists. Although the original outburst of radiation occurred at red-orange color frequencies more than 13.5 Bya, it appears today as a lower-frequency microwave radiation distributed nearly uniformly throughout the universe. The original radiation has been steadily red-shifted toward lower frequencies by the expanding universe and long ago passed out of the range of frequencies visible to the human eye and far-red frequencies that can be detected by the human skin as heat. The intensity of this radiation varies by only about one in one hundred thousand parts across the universe. This remarkable uniformity reveals that the distribution of atomic matter when this radiation was released from the prison of incessant scattering in the 400,000-year-old universe was nearly uniform. The extremely slight non-uniformity in the initial distribution of matter would influence the distribution of the first stars and provide a blueprint for the future evolution of the universe.

> *Or, taking man in his present position or consciousness, how or when may he be aware of that first cause moving within*

his realm of consciousness? In the beginning there was the force of attraction and the force that repelled. Hence, in man's consciousness he becomes aware of what is known as the atomic or cellular form of movement about which there becomes nebulous activity. And this is the lowest form (as man would designate) that's in active forces in his experience. Yet this very movement that separates the forces in atomic influence is the first cause, or the manifestation of that called God in the material plane! [Edgar Cayce reading 262-52]

God manifested himself in the early physical universe through the unified fundamental forces of nature and, from our perspective, especially through the electromagnetic force that guides and determines the atomic structures that form the natural elements, the basic building blocks of the universe. The universe is based on duality, a balanced state of tension between opposites, which is primarily expressed in nature as particles of opposite charge. The electric and magnetic fields that are associated with the electromagnetic force are intimately connected to each other and to electric charge, which can be positive or negative. A static charge produces an electric field, and a moving charge produces a magnetic field. A changing electric field produces a magnetic field, and a changing magnetic field produces an electric field. Electromagnetic energy such as light is an alternating electric and magnetic field that propagates at the speed of light in a vacuum (in the absence of matter) and can interact with electrons bound to an atom.

Each atomic force of a physical body is made up of its units of positive and negative forces, that brings it into a MATERIAL plane. These are of the ether, or atomic forces, being electrical in nature as they enter into a material basis, or become MATTER in its ability to take on [electrical attraction] or throw off [electrical repulsion]. [Edgar Cayce reading 281-3]

When the properties of light waves were being intensively studied in the 1800s, the method of propagation of light through space was not well understood. The propagation of sound waves through gases was understood, and it was natural to infer that light waves behaved in a manner similar to sound waves. Scientists theorize that there was an as yet undetected medium called the ether that acted something like the atmosphere in its role as a medium that supported the propagation of sound waves. Since no one had seen or measured the ether, the search was on to prove or disprove its existence. It was assumed that the ether moved freely throughout the universe or at least existed throughout the universe as a static field because light is able to travel to the earth from remote stars many light years distant from the earth. It was also thought that as the earth revolved around the sun, part of the year it would move in one direction relative to the ether, and six months later it would move in the opposite direction relative to the ether. Like sound waves traveling through the atmosphere, the speed of light and the frequency of light waves should be different depending on the direction the earth was moving with respect to the ether. Michelson and Morley devised an experiment to test this hypothesis and discovered that the speed of light was the same no matter what direction the earth was moving in around the sun. Their experiment, verified by later experiments, refuted the concept of ether as a substance through which electromagnetic radiation propagates.

The term "ether" as it appears in the readings does not seem to have the same meaning or the same inferred properties as the ether searched for during the Michelson-Morley experiment. It also carries somewhat different connotations in different readings. Several readings use the term "ethereal plane" to mean the spiritual realm (294-103, 900-362), which strongly indicates that the use of the term ether in the readings may be synonymous with spirit. Some mention an ethereal body as being synonymous with or related to an astral body (136-70) or a soul body (294-103). Others seem to use ether in a way that suggests it doesn't describe a specific property of matter or spirit, but is used in a more general sense to mean insubstantial or tenuous. Reading 254-68 uses the term in plural form "ethers of a material world" to perhaps suggest a property of the physical world upon which electromagnetic waves can be transmitted but also suggests that the ether is responsible for bringing positive and negative forces of the atom into the material plane. In this use ether appears to be the spiritual precursor of matter, a pre-electrical substance behind the sea of electromagnetic radiation and atomic matter that fills the universe, which we might assume is spirit or closely related to spirit. In this meaning, ether probably does not refer to a medium that is essential for the propagation of light through the universe but may be more akin to the current view of the vacuum energy of space, the medium out of which the source material of matter arose. The existence of atomic matter as a union of positive and negative electrical forces is mentioned in several readings and is consistent with the current scientific understanding of matter, but the implications of reading 281-3 goes beyond the scientific description of the nature of electrical charge. This reading suggests that the division of electrical force, or charge, into positive and negative aspects is somehow associated with the transformation of ether into the electrical activity that is the basis of materiality.

> *How can an article by mechanical means be charged with positive electricity continuously and economically? (A) It may not - without having its counterpart in the negative. For in manifestations, this is the basis of materiality. Only when they become spiritual are they ONLY positive. For materiality IS - or matter IS - that demonstration and manifestation of the units of positive and negative energy, or electricity, or God. [Edgar Cayce reading 412-9]*

Reading 412-9 reinforces the idea that charge separation or division of a proto-charge substance into dual and opposite aspects is a necessary condition for the formation of matter. Gamma-ray photons associated with the vacuum energy of the early universe could only decay into massless virtual particle pairs having opposite charge. They could not decay into two particles having the same charge. Conservation of charge remains a fundamental property of the universe today. A photon still cannot create two particles of the same charge but must create an electron-antielectron pair or a quark-antiquark pair with one of the pair of particles having a negative charge and the other particle having a positive charge of the same magnitude to conserve total charge.

At the atomic level, positive electrical charges are associated with protons that contribute to the nucleus and negative electrical charges are associated with electrons that coalesce around the nucleus. The position of each electron surrounding the nuclei cannot be known with certainty; it can only be described by a more nebulous probability wave function that can yield the likelihood that the electron can be found in a particular location. Unlike electrical charges attract each other, and like electrical charges repel each other, so the cloud of negatively charged electrons that surrounds the positive central core of an atom should collapse into the nucleus. The electrons in the cloud do not collapse and annihilate with the protons in the nucleus because their position and motion is constrained by quantum mechanical principles that cause the kinetic energy of an electron to increase as it approaches the nucleus (Heisenberg Uncertainty Principle) and prevents more than two electrons from occupying the same energy level in the electron cloud (Pauli Exclusion Principle). The effects of these two basic principles of quantum mechanics keep atoms from collapsing, give volume to atoms and the solid objects they form, and determine how atoms interact and exchange electrons to form different chemical compounds.

The main problem at this point in the evolution of the universe is that there are no heavy elements to use as building blocks or the components of more complex structures. For all practical purposes the only available elements are hydrogen and helium. The universe will need to build some heavy-element factories or forever remain a gigantic expanding gas cloud.

Stars and Galaxies

Phase 5: This phase begins at the creation of atoms about 400,000 years after the birth of the universe and brings us to the vast expanse of planets, stars, and galaxies we observe in the universe today, a long period of about 13.8 billion years. Only the three lightest elements of the ninety-two known natural elements in the universe were formed before the beginning of this phase. It took about 400,000 years after the formation of the universe to produce this first allotment of atomic matter, but this entire process whereby electrons and quarks precipitated into tightly bound units of energy we call atoms occurred relatively abruptly in cosmological terms. The formation of these first atoms released primordial photons from the bondage of incessant collisions with electrons and gave them the opportunity to ride the outward expansion of the spacetime structure while retaining a ghostly image of the distribution of matter in the universe at the moment they were set free. The nuclear reactions necessary to produce these elements were possible only because temperatures and pressures in the early universe were still large enough to support basic thermonuclear fusion reactions. As these new elements separated and cooled in the expanding universe, it was no longer possible for nuclear reactions to occur. The universe at the beginning of Phase 5 was composed almost entirely of three elements and contained about 75 percent hydrogen atoms, about 25 percent helium atoms, and a trace amount of lithium atoms.

The three new elements were too light to form any of the materials we recognize in our current environment today. It would take another 1.6 billion years, a long period during which the universe was dark and cold, before the hydrogen and helium could concentrate in isolated regions of space at the density and temperature required to reignite thermonuclear fusion reactions and begin the long process of forming the remaining elements. There was no light during this period except for the red-orange glow of the outward racing background radiation because there were no stars. Stars are atom factories, and without them the universe could not evolve beyond a vast region filled with gas. The heavy elements that comprise the universe we know today were created by nuclear synthesis in the high-temperature furnaces that exist in the interior of stars and that reproduce conditions similar to those that existed throughout the universe during the formation of the primordial hydrogen and helium nuclei. Two factors control the range of elements that are able to form in the interior of a star: the mass of the star and the amount of energy it takes to bind together the nucleus of each element. The relative abundance of the elements seen in the universe today was determined by the production of nuclei in stars.

The three new and lightest elements were nearly uniformly distributed throughout the expanding universe, but not quite. Their density distribution through space varied by only about one in one hundred thousand parts and that small non-homogeneity was sufficient to induce a small net gravitational attraction within slightly denser regions of atoms that existed within the surrounding slightly less dense regions. These regions of higher density were the seeds that caused the formation of local concentrations of atomic matter that were dense enough to collapse into isolated compact masses that could form stars. Concurrent with star formation, the global expansion of matter and energy throughout the large interstellar regions of the universe continued unabated. The following reading correctly describes the formation of stars from nebulae, the enormous conglomeration of gases and dust (bits of heavier-element debris from previous star explosions), and the fact that the process of star formation is universal, occurring constantly throughout the entire universe.

> *For, it is only reflected force that man may have upon those forces that show themselves in the activities, in whatever realm into which man may be delving in the moment - whether of the nebulae, the gaseous, or the elements that have gathered together in their activity throughout that man has chosen to call time or space. And becomes, in its very movement, of that of which the first cause takes thought in a finite existence or consciousness. [Edgar Cayce reading 262-52]*

However, there is one problem with this explanation as the process whereby the first stars were formed. The gravitational attraction of the light-element atomic matter within the higher-density seed regions was not strong enough to overcome the overall global expansion of the universe. To understand how stars were formed and where they were formed within this initial universe of distributed atomic matter, we must look again at inflation and introduce a new type of matter called dark matter, but for the moment we will assume that light-element atomic matter was capable of condensing into stars within these slightly denser regions of matter by itself. The structure of these denser regions took the form of weblike interlaced strands or filaments. The junctions of the major filaments commonly were the sites of the larger clumps of matter. The white, irregular spider webs that pop up on houses and trees during Halloween are a good depiction of the structure and distribution of these denser regions of space that were the birthing localities of the first stars.

When hydrogen and helium atoms are compressed by gravity into a compact mass at the core of a new star, the density, temperature, and pressure increase until the electrons are stripped from the atoms and the nuclei come together with sufficient force to fuse them. Thus, the core of a star mimics the high density and temperature conditions of the early Phase 3 universe. The stellar core pressures must be large enough to overcome the long-range repulsive electromagnetic force between positively charged protons. When the protons are forced close enough together to be within the sphere of influence of the short-range attractive strong nuclear force, they can fuse into heavier nuclei. Once a new, heavier nucleus is formed, the excess energy of fusion is released into the core of the star, the same process that man has learned to mimic in the hydrogen bomb. As long as there is fuel, the thermonuclear process will continue to create heavier nuclei from the original hydrogen and helium and from previously formed heavier nuclei. The amount of energy released during thermonuclear fusion is enormous. The energy migrates outward from the core of the sun and is released into space, mainly as X-ray radiation.

In the sequence of elements from hydrogen to iron, the nuclei of the heavier elements are more tightly bound than the nuclei of the lighter elements. Helium is a major exception and carbon and oxygen are minor exceptions to the rule. Stellar thermonuclear fusion can produce helium and the sequence of nuclei from carbon to iron but cannot produce nuclei heavier than iron. Not all stars can produce the full suite of nuclei from carbon through iron because some may not be massive enough to create the temperatures and pressures necessary to fuse the nuclei of heavier elements. Smaller-mass stars can only fuse hydrogen into helium, heavier stars can fuse elements as heavy as carbon or oxygen, but only the most massive stars can produce elements as heavy as iron by nuclear fusion. As long as thermonuclear fusion continues, the outward pressure of the escaping energy can balance the inward gravitational attraction of the matter within the star, but when the fuel gives out the star collapses and dies.

When it has exhausted its fuel at the end of its life, a less massive star will undergo a glorious last-gasp expansion into a red giant, then collapse into a compact white dwarf (perhaps the size of the earth's moon) surrounded by a planetary nebula, the remains of the outer portions of the star that are ejected and drift into space as the white dwarf forms. A larger and more massive star can have an even more spectacular end. It can form one or a series of several huge red supergiant phases almost as large as our solar system as it cycles through periods of expansion and collapse. During its final moments its earth-sized core will collapse into a neutron star (perhaps a few miles in diameter), and a massive explosion will drive the remainder of the star into space to form a supernova, a tremendous shock wave that carries large amounts of heavy elements into interstellar space. The explosion itself is so intense that it causes elements carried in the shock wave to fuse and form elements much heavier than iron. There is a second method by which elements heavier than iron can be formed. Two neutron stars can collide because of mutual gravitational attraction, causing individual neutrons to clump together and be ejected into space by the force of the collision. These neutron clumps can then transform into heavy element nuclei as some of the neutrons decay into protons by emitting electrons and antineutrinos to form the cores of stable heavy elements.

Because of irregularities in the distribution of mass, clusters of dust and gases rotate as they gravitationally condense into solar systems centered on a star. Similarly, when enormous clusters of stars gravitationally attract, they produce galaxies that rotate around a central black hole. Galaxies are formed where gravity acts on a denser regional distribution of stars, interstellar gases, and dust. As these materials become gravitationally bound and pulled toward their common gravitational center, there is also local attraction among the various objects, which creates non-uniform clumping. Gravity doesn't act equally across all parts of irregular objects so they are not all attracted toward a common point but move past the center of attraction and begin orbiting around it. This causes new galaxies to rotate and develop the flattened disk shape that is characteristic of most galaxies. Our sun, which resides in the outer edges of the Milky Way galaxy, completes one rotation around its center about every 225 million years and will complete about fifty revolutions over its life span.

Most large galaxies contain a supermassive black hole at their gravitational centers. Each black hole ranges in mass from millions to billions of times the mass of our sun and their gravitational attraction is so strong that they capture and incorporate into their cores interstellar gases, dust and even nearby suns and planets. There are an estimated 200 billion galaxies in the universe. Each one is anywhere from 3,000 to 300,000 light years in diameter and each are composed of about 100 billion stars. One light year is the distance light travels in one year or about six trillion miles. Galaxies ride the expanding spacetime continuum, moving away from each other at an average speed of about 45 miles per second per 19 trillion miles of separation. What appears visible to us in the night sky as a majestic slow-motion dance of celestial objects is actually a frantic outward race of stars and galaxies at unimaginable speeds. The size of galaxies in the universe and the number of stars they contain are almost too staggering for the human mind to comprehend (Oesch and others, 2016). In the far distant future 45 billion years from now, the spacetime fabric of the universe will have expanded another twenty powers of ten. There will be much more spacetime but the density of matter and energy in the universe will decrease tremendously and the universe will gradually go dark for any observer.

Many previously unknown planets, stars, galaxies, and other strange objects have been discovered throughout the vast universe in recent years as the quality of telescopes has improved with better mirrors and mirror arrays, computer algorithms to compensate for slight mechanical defects in the mirrors, and the development of solid-state digital detectors for the entire known range of the electromagnetic spectrum from low-frequency radio waves to ultrahigh-frequency gamma rays. Telescopes now sit in high orbit above the earth well outside of the atmosphere to capture incoming radiation before it interacts with and is distorted by atmospheric gases. Atmospheric effects such as the absorption of certain radiation frequencies and the scattering and blurring of incoming radiation limit the resolution of ground-based telescopes. The increased resolution and range at which radiation is now detected allows scientists to, among other things, measure orbital velocities of binary stars, to calculate minor variations in the light intensity from stars to indirectly identify the presence of planets, and to accurately measure the frequency of pulsating stars. The James Webb telescope that was launched into space in December 2021 is designed to detect objects a hundred times fainter than the faintest objects detectable by the Hubble Space telescope that was launched into space in 1990 and will see back in time to within a few hundred million years of the beginning of the universe, before the first stars were formed.

Dark Matter and Dark Energy

The universe as currently understood is composed of only about 4 percent atomic matter and trace amounts of electromagnetic radiation and neutrinos. Dark matter is postulated to constitute about 23 percent of the universe and dark energy the remaining 73 percent of the universe (these values differ slightly in the scientific literature), but the nature of these materials is not well understood and their very existence is inferred, not proven. The nature of these materials is not well understood because they have never been directly observed or experimentally detected. We only have indirect evidence of their existence based on the inconsistencies and discrepancies between the observed motions of large concentrations of atomic matter and their predicted and expected motions based on known and measurable gravitational effects. Their existence is inferred because our observation of the behavior of certain cosmological phenomena cannot be explained unless these materials are postulated to exist. Their names are place holders given by scientists to substances that they don't fully understand until they do have a better understanding of their true nature and properties.

The term "dark matter" is used to describe an essential component of the theoretical framework of cosmology that describes the formation, distribution, and evolution of clumped matter in the universe. Dark matter is "dark" because, unlike ordinary matter, it does not radiate, absorb, or emit measurable electromagnetic energy. Thus, it cannot be directly observed by our senses or astronomical instruments. However, dark matter interacts gravitationally with itself and atomic matter and the effect of this gravitational attraction can be measured. The initial distribution of matter in the early universe was quite uniform. Then how did the universe make the transition from a smooth, homogeneous distribution of matter and energy into the clumped matter that we observe as stars, nebulae, and galaxies? The irregular distribution of matter in the present-day universe has its origins in the inflationary expansion of the universe. The same rapid expansion that was reducing the overall density of the universe also greatly amplified minute vacuum fluctuations of the Higgs field. These vacuum fluctuations are thought to be the source of dark matter, although the details of that connection are yet to be confirmed (Chaves, 2019).

Every location in the universe has a sphere of influence called the horizon that sets a limit on the effective radius of the electromagnetic and gravitational interaction. This is true because both electromagnetic and gravitational waves propagate at the finite speed of light, the maximum speed at which information can be transferred through the universe. The influence of dark matter is affected by these horizons because the denser regions didn't expand as fast as the less dense regions, so the density contrast became magnified. During the inflation period, the roughness of the universe, a measure of these dark matter density variations, increased by twenty powers of ten, which meant that small higher-density seed regions vastly expanded into huge higher-density anomalies. Dark matter, which attracts other dark matter and atomic matter, provided the extra gravitational attraction that caused atomic matter to aggregate within these high-density dark matter anomalies and form the interlaced weblike distribution of stars and galaxies observed today in the universe.

Dark matter is postulated to have matter-like properties that allowed it to facilitate the formation of the first stars that lit up the early universe by causing the necessary extra gravitation attraction needed to pull shapeless clouds of light hydrogen and helium atoms into small compact masses. Not only was the extra gravitational attraction needed to overcome the universal expansion of atomic matter and cause it to collapse into stars, it was also needed to overcome the friction and outward radiation pressure produced as the inward-collapsing atomic matter heated. Dark matter is also necessary to fully explain the gravitational lensing of light by massive objects. Large concentrations of dark and atomic mass introduce curvature into the spacetime fabric, which causes massive objects like stars and galaxies to act like a lens and bend and focus light passing near the objects. The distortion and magnification of images of galaxies physically located in the background behind a gravitational lens relative to an observer on the earth cannot be explained only by the gravitational strength of the atomic matter contained within the lens (see cover image of a horseshoe Einstein ring). Dark matter is also needed to explain the behavior of rotating galaxies and galaxy clusters, physical situations that are impossible in the context of the known strength of atomic matter gravitational attraction. Scientists can calculate the net attractive force of gravity on a galaxy from the estimated mass of atomic matter within a galaxy. The calculated gravitational force in a rotating galaxy is usually not strong enough to keep the galaxy from shedding stars as it spins in space. Stars in the outer edges of galaxies are rotating faster than expected based on atomic matter gravitational attraction alone, which implies that there must be an additional source of gravitational force that holds each whirling galaxy together as a cohesive unit (Massey and others, 2010). Similarly, the calculated gravitational attraction of atomic matter within galactic clusters is insufficient to hold the clusters together and keep them rotating as a cohesive unit.

The term "dark energy" refers to an unknown and unobserved source of energy that is postulated to be present in the structure of the universe and to cause the observed present-day accelerating rate of outward expansion of the universe. As astronomical instruments were improved over the past decades, astronomers were able to observe and analyze the brightness and velocity of certain specific supernova explosions that were at various distances from the earth. These measurements show that the more distant supernovae are moving away from the earth at a greater velocity than nearby objects, and the same is true for other celestial objects that are more distant from the earth. The universe is falling outward toward a state of lower gravitational energy and is expanding at a constantly increasing velocity; that is, the rate of expansion is accelerating. Even as galaxies are being formed locally by the gravitational attraction of their components, individual galaxies across the universe are moving farther apart at a rate that increases over time. Astronomical measurements show the expansion of spacetime has continued since the origin of the universe. The rate of expansion was enormous during the inflationary phase, was much slower during the next 7.5 billion years, and since then has been steadily increasing. The current accelerating rate of expansion will continue forever. The known and estimated distribution of energy contained in the entire universe is not sufficient to prevent the attractive gravitational force of the dark matter and atomic matter it contains from pulling the universe back into a collapsing sphere. There is not enough energy to explain how the outward acceleration can continue. Dark energy is postulated to be the extra energy in the universe that keeps driving matter outward and is also postulated to be connected to vacuum fluctuations in the spacetime continuum.

It has been speculated that dark matter and dark energy are manifestations of the activity of spirit that is just beyond our capacity to detect or view with present-day scientific instruments. Are they relatively unmodified forms of spirit that pervade the physical universe and control its local contraction and global expansion so that atomic matter and energy would have the greatest potential to form a planet capable of supporting biological life? If we are able to detect them in the future, might we be able to prove that God exists or would we only understand them as extensions of the physical universe? As presently postulated or understood, both dark matter and dark energy are related to the vacuum fluctuations of the Higgs field that underpins the spacetime continuum, so they are not thought to be some new etheric materials but only aspects of the spacetime fabric that influence the behavior of atomic matter. It is unlikely that any scientific evidence that proves the existence of these illusive and speculative forms of matter and energy, or better defines what they really are or what they actually represent, will be accepted by scientists as proof that God exists or evidence of a God material or spiritual substance.

If the vastness of the universe; its origin as an unimaginably small bundle of energy undergoing an enormously rapid expansion; the birth of atomic matter, stars, and galaxies; its ongoing destruction as pieces are captured and destroyed in the cores of black holes; and the eventual dispersal of the remainder of the universe as a massive dark black void isn't enough to make someone think there might be more to the universe than happenstance, then why should the discovery of another piece of the puzzle change anyone's mind? The real problem is that we already see manifestations of God's creative activity every day and don't recognize them for such. Although science keeps discovering more and more about God's creation, it will require a change in consciousness for man to realize and understand that we are just uncovering a few more examples of the infinite expressions of God.

The Solar System

Think of it this way. God's children have become severely addicted to the drugs of selfishness and self-gratification. In the pursuit of their selfish ecstasy, they bring discord and disharmony into the spiritual realm without realizing that their mental state and activities are detrimental to their very existence. They have catered to their selfish and spiritually rebellious desires for so long that they have forgotten their spiritual ancestry and the special relationship they once had with God. They no longer remember who they really are or the glorious spiritual heritage they could have claimed. God could have decided to write them off and let them follow their destructive path to its inevitable conclusion, but in his compassion and love for them he decided to build a rehabilitation facility that will help them recover from their addiction, remember their lost estate and reawaken their buried, but ever-present, innate desire for a loving relationship and companionship with their Father. Through activity of Mind, God unleashed a new universe in the form of spirit/energy expressed as the vacuum energy of a spacetime continuum. He caused spacetime to expand at a phenomenal rate, enormously increasing the size of the universe, and forcing it to undergo a phase change whereby a substantial portion of its energy would clump together as electrically charged particles called quarks and electrons and uncharged neutrinos nearly uniformly scattered across the universe. Quark triplets self-organized into positively charged protons and neutral neutrons, which combined to form large amounts of hydrogen and helium nuclei and trace amounts of lithium nuclei, the cores of the three lightest elements. All of this took place in only about sixteen minutes as human beings measure time. Temperatures were still too large for the free electrons to combine with the nuclei. It would take another 400,000 years before the universe would cool enough to allow the free electrons to bind to nuclei and form atomic matter and release the electromagnetic radiation that had been trapped by collisions with the free electrons.

Not until about 1.6 billion years after the universe began would the gravitational attraction of the slightly denser regions of atomic matter and dark matter overcome the overall expansion of the universe and cause large regions of atomic hydrogen and helium to be pulled together and collapse into stars. Inside these dense concentrations of matter, the nuclei of the atoms were stripped of their electrons and subjected to sufficient pressure and temperature to cause thermonuclear ignition, which fused the light nuclei into more massive nuclei of heavier elements. Points of light began to appear all across the universe. As the stars burned the last of their fuel and died, these new heaver nuclei acquired electrons from the outer regions of the star to form new elements as they were broadcast throughout interstellar space by explosive dispersal. These new elements mixed with the original hydrogen and helium in interstellar space and elements ejected from neighboring stars and became the raw material for new stars. The birth and death of a star is one single cycle in the continuing process of heavy element formation in the universe. The cycles of solar life and death still drive the creation of new atomic matter 13.8 billion years into the life of the universe and spread the new matter through interstellar space but on a smaller scale than before because the continuing expansion of the overall universe is weakening the gravitational attraction that causes interstellar matter to coalesce. It took 9.3 billion years before our sun and its associated planets on the outskirts of a rather ordinary group of stars called a spiral galaxy caught God's attention as a potential habitat for humanity, and it took an additional 4.5 billion years for the earth to mature and become habitable for the plant, animal, and human life we see today. The patience of Job is nothing compared to the patience of God.

The gravitational attraction that brings interstellar gas and dust together to ignite thermonuclear fusion and form suns is the same force that creates planets. When a cloud of interstellar dust is pulled together into a star, density variations in the collapsing cloud cause the entire cloud to rotate around the future star, the center of the gravitational attraction. The effects of density variations and rotation also cause the dust cloud to agglomerate into smaller clumps of matter that become planets that orbit the new sun and that have their own unique density and rotate about their own axes. The masses of planets are far less than the masses of suns. There is sufficient mass and enough internal friction generated during the collapse and compaction of planets to generate high temperatures in their cores but not enough to ignite thermonuclear fusion. In our solar system, debris from the interstellar dust also formed a large disk of smaller ice objects (primarily methane, ammonia, and water) called the Kuiper Belt that extends outward from the orbit of Neptune to a distance equal to about two-thirds the radius of the solar system. An even larger spherical cloud of dust and billions of comets three light-years in radius called the Oort Cloud surrounds the solar system. The phrase "mists that are gathering" that is used in reading 900340 would seem to well describe this drawing together of interstellar gases and dust to form stars and planets.

> *... we find first, as the worlds created - and are still in creation in this heterogeneous mass as is called the outer sphere, or those portions as man looks up to in space, the mists that are gathering - what's the beginning of this? In this same beginning, so began the earth's sphere. [Edgar Cayce reading 900-340]*

Each sun we observe was formed as an integral part of a larger solar system that can contain one or more planets and smaller lumps of rock and ice. There are trillions of suns in the universe, so the probability of finding more earth-size planets in favorable orbit around a sun that emits an electromagnetic spectrum favorable to carbon-based biological life is large. Planets that form closer to the sun typically contain heavier elements and are rocky, whereas planets that develop farther from the sun usually are composed of lighter elements and are gaseous, although these gases are usually in frozen form on the surface because the planets are distant from the sun and are quite cold. Planets can gravitationally trap smaller bodies of rock and ice, called moons, which orbit the planets. It is likely that the distance of the earth from the sun and its orbital period were important factors to consider in its choice as a future dwelling place of carbon-based biological life and souls. These factors largely determine whether the planet can maintain a reasonable average temperature, an atmosphere of the appropriate density and chemical composition, and water bodies that would support the presence of complex biological life. Usually the inner planets of a solar system are much too hot and the outer planets are much too cold to support biological life as we know it.

Planets that orbit other suns (exoplanets) are extremely difficult to detect because of their large distance from the earth, small size relative to other objects in the universe, and because these small dim objects are found so close to stars that are extremely bright as compared to the small amount of reflected light coming from the planets. However, astronomers have made vast improvements in the size, sensitivity, and detection spectrum of telescopes and have placed telescopes in space where atmospheric distortion and absorption of targeted radiation has been eliminated. Because of these advancements, scientists have been successful in their quest to find unknown planets, and the search continues with the anticipation of finding an earth-like planet that might possibly support carbon-based life (Vanderburg and others, 2020 and summary article). Nearly 5,000 exoplanets, most having a radius of from one to five times the radius of the earth, have been confirmed, and more than that are candidate planets awaiting confirmation. By far, most of these exoplanets were confirmed by detecting a variation in the brightness of a star as the orbiting planet moved in transit across the face of the star and blocked a small portion of the starlight coming toward the earth (NASA Exoplanet Archive). Only about sixty exoplanets have been directly detected by telescopes, primarily by the Hubble telescope in orbit around the earth, and a few by linked ground-based telescopes that have a much higher resolution than single-lens telescopes. Linked telescopes are better at finding large diameter planets orbiting far from their suns, like Jupiter in our solar system.

Despite the technological difficulties, the quest for extraterrestrial life is underway with radio telescopes scanning the sky listening for evidence of communication by nonhuman civilizations. Signals have been transmitted into the vastness of space and are now radiating away from the earth at the speed of light in the hope that someday somewhere they will be detected by another civilization that will respond by sending an answering signal back toward the earth. The attempt by astronomers to locate extraterrestrial life is predicated on the idea that biological life can arise spontaneously under the right planetary conditions. Because there are about 200 billion galaxies in the universe with each having about 100 billion stars, the odds of finding an earth-like planet in a solar system associated with one of those stars is considered highly likely. Does that mean there are more earth-type planets supporting life scattered across the universe. Not necessarily. The readings do not support the existence of an extraterrestrial species of biological life like the lower life forms of the earth, nor do they suggest the presence of any other-worldly race of humanlike or nonhuman people. They suggest that the attempt to communicate with extraterrestrial life will end in failure because man is the only sentient species in the universe and is here as part of a specific plan set in motion by God for the purpose of soul development. Two readings clearly state that no other humanlike life or any form of animal life will be found on any other planet. This doesn't rule out lower plantlike life forms or the possibility of other types of life, but that may only be because the questions that were addressed in the reading specifically inquired about human and animal life.

> *Are any of the planets, other than the earth, inhabited by human beings or animal life of any kind? (A) No. [Edgar Cayce reading 3744-4]*

> *Upon what planets other than the earth does human life exist? (A) None as human life in the earth. [Edgar Cayce reading 826-8]*

The readings endorse a version of astrology that has little in common with the use of astrology to provide practical advice for readers of newspaper columns. The readings assert that certain celestial bodies, including the planets of our solar system, are associated with soul journeys into dimensions of consciousness that the soul needs to experience because of a deficiency or shortcoming in its spiritual development (3349-1 and 5177-1). These periods of residence in nonphysical dimensions of consciousness between earthly incarnations allow the soul mind to receive the specific spiritual training needed to purge it of certain selfish traits. These dimensions of consciousness are also described as existing in higher vibrational states than the physical space dimensions. The higher vibrational states are not detectable using current scientific instruments but memories of those experiences are sometimes discernible to the minds of perceptive individuals. Cycles of incarnations on the earth give souls the opportunity to make practical physical application of the mental training received in these dimensions of consciousness and provide an opportunity for souls to demonstrate that they have mastered, or failed to master, the celestial lessons.

The relative positions of the planets, sun, and earth's moon of our solar system and certain stars appear to influence the personality of an incarnated soul because the soul consciousness completed training in a nonphysical realm associated with the particular celestial object just before it entered the earth. These influences are felt in the human consciousness of some individuals as latent urges or compulsions. They seem to correlate with the astrological signs at the birth of the soul because of the recent soul activity, not because of any physical property of the celestial objects. These urges are not as a directive, instruction, or mandate to behave in one manner or another but usually are as subtle remnant memories and mental influences. The strength of the influence, the force of their effect on the individual soul mind, and the intensity of the residual feeling that might guide an incarnated soul to follow the urges through directed physical activity are determined by the application of the will of the soul (Willner, 2001). The force of the will of the soul is always greater than the force of the latent soul memory.

> *Remember, in the analysis of the urges which are latent and manifested, which arise from activities in the earth or the realms through which the entity passed, none of these urges surpass the will of the individual entity; that which is the gift of the Creator, that ye may make yourself one with the Creative Forces, and thus indeed a child of the living God. [Edgar Cayce reading 5254-1]*

The readings make it clear that the planetary dimensions they describe are not physical dimensions but are states of consciousness that can be correlated with the physical planets. There are certain adverse mental states or conditions that can be associated with each planetary dimension and that can be corrected by periods of existence, training, and conscious awareness in these dimensions. This is not to mean that souls reside on the surface of other planets as we perceive them from a three-dimensional perspective but only that there is a nonphysical connection or association between these alternate dimensions and certain planetary objects (2753-2 and 2405-1). Neither physical beings nor nonphysical beings are walking around on the planets of our solar system. Planets are the physical analogs of the "locations" where souls get spiritual training and strengthen their minds to overcome spiritual deficiencies that are causing the soul to separate in consciousness from God. They may be akin to the "many mansions" referred to by Jesus (John 14:2), a term that is clarified in the readings to mean many states of consciousness. The readings indicate that there are millions of combinations of experiences necessary to address the myriad of problems and deficiencies of every soul.

> *This is more clearly demonstrated and interpreted in the words of the Master himself, "In my Father's house are many mansions", many consciousnesses, many stages of enfoldment, of unfoldment, of blessings, of sources. And yet God has not willed that any soul should perish, but has with every temptation, every trial, prepared a way of escape - or a way to*

*meet same; which is indicated here by the Creator, the Maker
of heaven and earth and all that in them be. [Edgar Cayce
reading 2879-1]*

Planetary dimensions are part of a greater complex of created dimensions that are being used as opportunities for souls to discover their true nature and to release them from their propensity toward selfishness and self-gratification. Because these dimensions, as given by the readings, mostly exceed the three familiar spatial dimensions of the earth, it might be tempting to correlate them with the multiple dimensions of superstring theory. Superstring theory is an attempt to explain all particles and forces in one theory (see the Mathematical Descriptions of the Universe section). It is expressed in nine or ten spacelike dimensions that might suggest a connection among the planetary dimensions of consciousness described in the readings and the nine physical planets (treating Pluto as a planet) or nine planets plus our sun. However, it is more complicated than it appears, and there is no clear or logical way to relate the planetary dimensions of the readings to the higher spatial dimensions of superstring theory. Five readings (311-2, 2753-2, 3006-1, 3188-1, and 5002-1) give dimensions associated with planets, but the dimensions are not unique. Mercury has been assigned dimensions 2, 5 (twice), and 7; Venus has been assigned 4 (twice) and 7; Mars has been assigned 1; Jupiter has been assigned 4 (twice), 5 and 7; Saturn has been assigned 1; and Uranus has been assigned 2, 6, and 8 (twice). Neptune and the Kuiper Belt object Pluto have not been assigned dimensions. The reason for the multiple and duplicate dimensional assignments to each planet in the readings is confusing and incomprehensible, and definitely means that they cannot have any relation to the nine- or ten-dimensional spatial framework of superstring theory. Neither should one expect a correlation between the two because the readings emphasis that the dimensions it describes are dimensions of consciousness, not higher spatial dimensions.

Strangely enough, the possibility of a multiverse allows us to take another look at the concept of nonphysical dimensions associated with the physical planets of our solar system. While our universe evolved in a way that permitted carbon-based biological life to thrive within the confines of a planet with specific biologically favorable surface conditions—such as a fairly narrow range of warm temperatures, abundant quantities of liquid water, and an appropriate atmospheric composition—the material composition of other universes may not be favorable for life as we know it. There would presumably be the equivalent of the fundamental particles and forces that we observe because the source material would be the same spirit/ energy out of which our universe developed, but the suite of fundamental particles and nature of the forces may look entirely different and may form an arrangement of atomic structures that bear no relation to those with which we are familiar. Whether or not life as we know it, or don't know it, could form in such strange environments is impossible to evaluate. The mathematical predictions that arise out of the multiverse theories cannot be verified by observation or measurement with instruments. These other universes would be out of the reach of our physical instruments, so we may never know the truth about the existence of such a plethora of universes, but the concept is intriguing and has been the topic of much speculation and investigation among physicists. It is unlikely that we will ever be able to detect or communicate with another member of a multiverse community because there would be no physical connection remaining between our universe and our sibling universes, but that doesn't preclude a spiritual connection.

If the soul training facilities associated with other dimensions of consciousness are instances of other universes, would nonphysical spiritual creatures such as souls have the ability to cross the divide that separates the multiple universes? Perhaps the physical planets are in some way associated with portals to these alternate universes, and the description of planet-associated dimensionalities used in the readings is meant to convey this multiverse concept. There is a possible, but tenuous, connection between the concept of multiverse and the readings. Some readings state that universes, plural, were created and reading 5755-2 elaborates that there is a great need for many different types of soul experiences because there are millions of souls, each with their own idiosyncrasies that need individual attention. Perhaps other universes with different causality conditions or other characteristics we cannot even image have been created and are also being used as centers for soul development.

GOD, the Father, then, is the Creator - the beginning and the end. In HIM is the understanding, BY and through those influences that have taken form - in universes - to meet the needs of each soul - that we might find our way to Him. [Edgar Cayce reading 5755-2]

... though there may be worlds, many universes, even much as to solar systems, greater than our own that we enjoy in the present, this earthly experience on this earth is a mere speck when considered even with our own solar system. Yet the soul of man, thy soul, encompasses ALL in this solar system or in others. [Edgar Cayce reading 5755-2]

The Earth and Life

First that of a mass, which there arose the mist, and then the rising of same with light breaking OVER that as it SETTLED itself, as a companion of those in the universe, as it began its NATURAL (or now natural) rotations, with the varied effects UPON the various portions of same, as it slowly - and is slowly - receding or gathering closer to the sun, from which it receives its impetus for the awakening of the elements that give life itself, by radiation of like elements from that which it receives from the sun. [Edgar Cayce reading 364-6]

When the earth became a dwelling place for matter, when [interstellar?] gases formed into those things that man sees in nature and in activity about him, then matter began its ascent in the various forms of physical evolution - in the MIND of God! [Edgar Cayce reading 262-99]

The readings state that our earth was chosen as the place for man's indwelling at some point in its evolutionary cycle (262-119) and that the Spirit and Mind of God acted on the earth to form the mineral, plant, and animal kingdoms that would ultimately evolve into the kind of environment necessary to support human bodies. Gravity had caused the slightly denser regions in the early hydrogen and helium matter distribution to clump and initiate thermonuclear reactions that could produce heavy elements. Repeated cycles of star formation and explosive star deaths created nebulae and interstellar matter enriched with these more massive elements. Galaxies and solar systems containing heavy-element planets precipitated out of the enriched interstellar gases and dust. Some planets had orbital characteristics that caused the temperature of the cooling planet to stabilize within a range favorable for sustaining carbon-based organic life. One rather obscure ball of magma and rock with an iron core, the third planet from the sun in a solar system of four heavy-element rock planets and four light-element gaseous planets circling an orange-red sun at the outer edge of an unremarkable spiral galaxy looked like it might support biological life.

The earth underwent many upheavals and changes as it cooled and took on its current spherical shape with a 4,000-mile radius. Today, the center of the earth is almost entirely composed of a 2,200mile radius sphere of iron and nickel consisting of a smaller solid core surrounded by a thicker molten shell. The interior is still molten despite the fact that the planet has radiated excess heat for about 4.5 billion years. An eighteen hundred-mile thick viscous liquid mantle surrounds the core and is covered by a three-mile to forty-five-mile thick solid crust. Various minerals that are lighter than the central iron-nickel core are present in the mantle and crustal rocks. The mantle is primarily composed of oxygen, silica, magnesium, and iron. The crust is primarily composed of oxygen, silica, aluminum, and iron. Continents are large masses of crustal material that migrate across the surface of the underlying magma under the influence and control of thermal convection currents in the magma, a process described by plate tectonics. Continents are created along lines of upwelling magma where molten rock pushes through thin areas of the crust, typically as long linear features stretching across ocean floors. The rising molten magma divides, cools, and migrates laterally in opposite directions, where it solidifies to form new crustal material. Continents are destroyed when one crustal plate collides with and slides beneath a neighboring crustal plate. The subsiding plate melts as it is forced into the underlying liquid magma. The overlying plate usually buckles and folds to create mountains and volcanoes. Linear regions along the surface of the earth where this geologic process occurs are called subduction zones. One of the compounds that formed as the earth cooled was water, a light slightly dipolar molecule that could exist in a liquid state as cooling progressed. Water, an essential ingredient for biological life, consolidated into seas and oceans that separated from and floated on top of the crustal mass, while water vapor and the lightest gaseous compounds floated on top of both oceans and continents to form an atmosphere around the earth.

> *But there went up a mist from the earth, and watered the whole face of the ground. (Genesis 2:6)*

Minerals are chemical compounds that have a specific chemical composition and an orderly internal structure and crystalline form. Recent scientific studies of different minerals and the environments that produced them have expanded the knowledge of minerals and highlighted their variety and complexity. Minerals started forming throughout the universe long before the earth was formed. These first microscopic minerals were part of the interstellar dust that came from the death of stars. About sixty minerals and mineral phases (minerals having the same chemistry but different structure) are found today in unmodified stony meteorites, parts of asteroids that assembled out of interstellar dust as our solar system formed. Because they have not been subjected to melting and cooling or high pressures, they retain their original composition. Stony meteorites that have been subjected to elevated temperatures and pressures can contain as many as 250 minerals.

Each new environment is a breeding ground for new minerals and mineral phases. Like other planets, the earth captured a wide variety of elements from interstellar dust during its formation. Since then, elements have been combining and recombining into new chemical compounds by exchanging and sharing electrons to arrive at more stable configurations depending on environmental conditions and geologic processes. Various and different minerals have been formed, and are still being formed, in the upper regions of the high pressure and high temperature environment of the mantle, in the lower pressure and cooler regions of the crust, in the solute-rich hydrothermal fluids that migrate through the subsurface regions of the crust, in the water of seas and oceans by the precipitation of minerals, and at the interface between crust and atmosphere in playa lakes and tidal mudflats where evaporative processes rearranged dissolved soluble elements into various minerals.

About 1,500 different minerals have been identified as geologic minerals. But the bulk of the minerals that presently exist on the earth do not have their origins in the geologic processes that give us the most memorable examples, such as massive quartz crystals or semi-precious and precious stones. Most of the mineral species on the earth were created after the first life form appeared. They are biological in origin and account for a majority of the 4,300 mineral species currently identified. These minerals are largely associated with plants, the soils and microbes that keep them healthy, and animal biological functions. Minerals have historically been studied in terms of chemical classification and crystal structure schemes but are now being studied in the context of mineral evolution, a progression of increasingly more complex compounds with different functions. Minerals are seen to evolve as interstellar dust becomes part of the planetary accretion of matter and is reformed by planetary geological and biological processes such as mantle cooling, crustal formation, compression pressure during continental movement, and the activity of biological organisms (Hazen and others, 2009). Science classifies the material of the mineral kingdom as inanimate matter, but the readings indicate that minerals have a spiritual component beyond the fact that all atomic matter is a form of condensed spirit. Besides being the substance of which the entire universe was created, Spirit also appears to be involved in ordering and organizing matter, perhaps acting locally in opposition to entropy, the tendency of a system to move from a state of order to a state of disorder.

> *In the beginning, when chaos existed in the creating of the earth, the Spirit of God moved over the face of same and out of chaos came the world—with its beauty in natural form, or in nature. [Edgar Cayce reading 3976-8]*

When the earth had cooled sufficiently and developed favorable average surface temperatures, oceans, a basic hydrologic cycle, an atmosphere, and complex inorganic and organic molecules it was chosen as the planet to receive the impetus that initiated biological life. Through the direct action and guidance of Mind and Spirit, rudimentary organic molecules were stimulated to create the first living organisms and start the process of natural biological evolution toward higher-order organic life. God was bringing order out of chaos and preparing an environment that would best present lost souls the opportunity to come face-to-face with their shortcomings and failures. Various stages of evolutionary progress altered and diversified the earth's life for billions of years and eventually caused it to become dominated by the plant and animal life forms that we see spread across the earth. Science has concluded that the living organisms we observe on the earth today began with single-celled organisms about 3.8 Bya. It cannot say that these first cells were created by God because that is a non-testable hypothesis not subject to analysis by the scientific method. It also cannot say how these first primitive cells originated or how complex biological molecules were first formed and how they might have self-organized into living cells.

The evolutionary force of natural selection can change the biological features and behavior of existing life, but evolution cannot explain how life might have arisen out of inert lifeless elements. The 1952 Miller-

Urey experiment, which simulated expected environmental conditions about 4 Bya, showed that over time, a heated water bath of simple organic molecules (water, methane, ammonia, and hydrogen) subjected to electrical discharges can form basic amino acids, which are the building blocks of proteins. Proteins are essential components of biological life forms. Since oxygen destroys simple amino acids, it is thought that the production of these compounds would have happened before the earth's atmosphere gained a significant concentration of oxygen. Later, similar experiments produced other amino acids, many of which are not used in protein synthesis. But the creation of these basic biological units from a soup of basic chemicals is a long way from the organized purpose and activity of billions of atoms and molecules in a living cell. Complex biological molecules are too complex to be created by chance.

> *... as this evolution [earth] began in the mind of the Creator, there then came the point, the place, the beginning, when that as created was GIVEN that as was necessary to make its development by applying these same forces ONE with that Creative Energy. [Edgar Cayce reading 900-340]*

Biological life did not begin because there was a fortuitous combination of organic molecules, water, and generally favorable environmental conditions. The earth's environment may have contained a few natural amino acids but these compounds were not arranged into groups of organic molecules that could self-organize into functioning cells that were capable of extracting energy from their environment for sustenance and reproduction. There was no sense of self-awareness and coordinated activity among these separate compounds. The Mind of God gave the impetus and rudimentary self-awareness that caused organic molecules to work in cooperation as the first unified organisms. Cooperation within and among cells is an essential element of healthy biological activity. The first lesson in the Search for God readings is cooperation, an indication of its importance in the education and spiritual growth of the soul. From the point of view of the readings, souls are corpuscles (cells) in the body of God. Cooperation between the soul and its creator and cooperation among individual souls is essential for harmony and peace in the spiritual realm. The act of biological creativity was an intentional interference with the physical evolution of the universe through the activity of the Mind and Spirit of God. Biological life was initiated so that over time, through the process of biological evolution, an environment capable of supporting higher-order beings would develop from physical matter. Spirit and mind were necessary to activate biological life from inert matter and are still necessary to animate physical matter. Science has not yet discovered this aspect of biological life because its attention is directed toward understanding life as matter and energy reactions and interactions.

Thus as has been indicated, the Spirit pushed into matter -and became what we see in our three-dimensional world as the kingdoms of the earth; the mineral, the vegetable, the animal - a three-dimensional world. [Edgar Cayce reading 262-114]

And this lesson should be gained: LIFE is action, and is made of the MENTAL action or mind. [Edgar Cayce reading 294-70]

The force of mind manifests spirit as the movement and activity of biological life as seen in the instinctual activity of animals and the will-directed activity of man. The biological life that arose on the earth is here for a specific purpose having to do with the creative expression and spiritual growth of souls. Did God tweak the biological evolutionary process at critical junctions to enhance the probability of certain outcomes that would lead to a human life form? Science has not detected or confirmed any specific instances of direct external interference but, as we shall see, the readings indicate that the modern human body is not simply the evolutionary endpoint of one of many random life forms that spontaneously appeared in a carbon-nitrogen-oxygen rich soup.

Matter is condensed Spirit;
Life is matter activated by Mind;
Humanity is life directed by Will.

Primitive single-celled creatures were the sole form of life on the earth for three billion years, during which time they flourished and produced a huge variety of forms as they evolved and explored different methods of extracting energy from their environment. Each cell contained a soup of ingredients needed to sustain life but had no well-defined internal structure. To be successful as a new life form, each variety had to encode within its structure the ability to reproduce, to direct the synthesis of new copies of itself so that it could survive as a species and engage in the evolutionary process. To accomplish this activity, cells had to build the two basic nucleic acid molecules (DNA and RNA) and use them to construct proteins. Only nucleic acids are capable of self-replication. DNA (deoxyribonucleic acid) holds the instructions for replicating the organism and producing RNA (ribonucleic acid), which contain the templates needed to build the amino acid sequences used in protein synthesis. These are the necessary and sufficient ingredients for the construction of the physical components of all of earth's biological life forms. The primitive single-celled organisms were generally microscopic and absorbed nutrients and released excrement directly through their enclosing membrane. The scientific theories of their origin differ. Perhaps the major contender among these theories is that they first formed near magma vents along mid-ocean ridges where temperatures were warm and stable and there was an accumulation of nutrients that provided favorable conditions for life to begin and survive. These marine cells are believed to have migrated to the surface of the oceans and later to coastal areas and dry land during their evolutionary history.

> *Yet there are those things that make for harmony in their relationships as one to another, as do the turmoils of the mother-water that brings forth in its activity about the earth those tiny creatures that in their beginnings make for the establishing of that which is the foundation of much of those in materiality. [Edgar Cayce reading 694-2]*

Other theories suggest a terrestrial origin for cells which was followed by their migration into oceans. In either case, they eventually developed the ability to extract energy from the early carbon-dioxide rich atmosphere through the process of photosynthesis and released oxygen into the atmosphere as a byproduct. They are considered to be the first large-scale source of oxygen to the atmosphere. The oldest fossils on record (stromatolites) were marine colonies of these algae-like cyanobacteria. By 2.5 Bya they had collectively released enough oxygen in the atmosphere to cause rust to form on iron-rich bands of rock, the first evidence that the atmosphere was on its way to supporting higher life that uses the oxidation of sugar and fats as an energy source instead of direct solar radiation. The demise of these first single-celled organisms might be an object lesson for man. They appear to have basically drowned in their own excrement; as oxygen concentrations rose in the atmosphere, the higher concentration of oxygen is thought to have killed off many of them and contributed to their decline as a species.

... for remember, each cell in an organism is as a universe in itself, each attempting to manifest the purpose for which its functions are set in an organism, and should coordinate one with another ... . [Edgar Cayce reading 433-1]

Single-celled organisms eventually evolved into more complex multi-celled organisms with special internal structures called organelles (rudimentary organs) that developed to handle specific biological processes such as photosynthesis, energy production from sugar, and cell membrane construction. They also formed a central nucleus designed to hold and better protect the cell DNA (Margulis and Sagan, 1997). The nucleus of every cell in a physical organism contains the same DNA and therefore partakes of the common purpose to express the basic characteristics of that organism, no matter whether that organism is a microscopic bacterium or a complex human being. When cellular matter is spiritualized, it recognizes the purpose assigned to it, a directive that it strives to fulfill while fitting harmoniously within the larger body of which it is a part and within its environment.

... the human body made up of electronic vibration, with

> *each atom and element of the body, each organ and organism*
> *of same, having its electronic or unit of vibration necessary for*
> *the sustenance of, and equilibrium in, that particular organ-*
> *ism. Each unit, then, being a cell or a unit of life in itself, with*
> *its capacity of reproducing itself by the first form or law as*
> *is known of reproduction, by division of same. [Edgar Cayce*
> *reading 1800-4]*

It took the universe about 9.3 billion years to evolve from a collection of fundamental particles and energy into suns and solar systems, one of which contained the planet that God chose as a rehabilitation center for souls. It took another 4.5 billion years for that planet to evolve a natural environment of biological life that could properly support the nutrient requirements of a human body designed as the ideal vehicle for the spiritual growth of incarnated souls. The Father of souls, the Father of the universe, and the Father of biological life is infinitely patient, knowing that all things come to maturity in their proper time and that the construction of this rehab facility would be well worth the wait. Besides, 13.8 billion earth years is a minuscule period in the life of his soul children. The pattern of evolution of the biological forms on the earth is similar to the pattern of evolution of the universe as a whole. It started with an initial seed or motivating influence, an organizing principle that was set in motion and then allowed to follow its path to maturity.

The creation and evolution of the physical and biological universe can be viewed as the sequential activity of spirit, mind, and will. Spirit is the basis and precursor of electromagnetic energy and the substance out of which subatomic matter particles of various masses condensed with the intrinsic property of charge duality. The expanding spacetime continuum, the attendant division of forces and subsequent decrease in temperature, pressure, and density brought about the conditions needed for the clumping of subatomic particles into nuclei and atoms. Stellar cores provided the environment that caused the most basic elements to be reassembled into heavier elements that could consolidate into large collections of inanimate matter. Certain atoms and molecules became organized into regular arrays as minerals.

When specific conditions were met, mind and spirit were allowed to influence the material forms of condensed spirit to give an organizing and cooperative component to certain classes of molecules. Groups of organic molecules with biologic potential began to work together to extract energy from the environment for the collective good of the whole organism. The whole became greater than the sum of its parts, and biological life proliferated. Mind, the builder, allowed rudimentary organisms to evolve into more complex organisms within the common context of developing new and better ways of extracting energy from the environment for sustenance while preserving the physical integrity of the organism and successfully propagating the organism in a highly competitive environment. This act of molecular and cellular cooperation is essential for the growth of all biological life just as mental cooperation between the incarnated soul and God is essential for soul growth. Life forms on the earth are still evolving and forming biological diversity just as planned. The earth is the crown jewel in the evolution of the universe; a blue and brown ball of water and rock surrounded by a shimmering atmosphere, illuminated by the fierce fires of a thermonuclear furnace on one side and swallowed up in the pitch-black darkness of cold space on the other side. By its daily twenty-four-hour rotation period, it absorbs and radiates heat in the right amounts to keep the average temperature over most of its surface within an acceptable range for biological life.

The oceans of the earth were already forming by about 4.4 Bya. Life probably began in the oceans as primitive single-celled life forms some 3.8 Bya to 4.0 Bya. It wasn't until about 2.1 Bya that the first multi-celled organisms took hold and organisms with more cells and greater complexity started developing. Cells and organisms developed that began to metabolize the large reservoir of oxygen that had filled the atmosphere and used sexual reproduction to create offspring. The first rudimentary land plants didn't appear until about 1 Bya but by about

500 Mya (million years ago) the first terrestrial and marine animals were forming. The dinosaurs did not appear until about 230 Mya and the first mammals appeared about 130 Mya. The great apes (orang-utans, gorillas, and chimpanzees) have only been separate species since about 15 Mya, and it wasn't until about 7 Mya that the chimpanzee and earliest humanlike species parted ways to take separate evolutionary paths. Many of the rebellious souls who were creating trouble in the spiritual realm were watching the evolution of earth's biological life and were imagining ways to use the new physical life forms to gratify their selfish desires.

Before the current human body was made available for incarnation, souls gained control over certain forms of biological life when they used their will to forcibly insert themselves into some lower life forms and probably into the most highly evolved life forms, the naturally evolved Homo sapiens species. They saw the earth as an opportunity to experience carnal sensations, which they satisfied by pushing into the earth's environment where they modified and interfered with the natural evolutionary progression of earthly creatures before the earth was fully prepared and ready for their entry for the purpose of soul rehabilitation. In the process they became enamored with the sensual attractions of biological life and accelerated the ongoing loss of conscious awareness of their true relationship with God. To counter this unacceptable and chaotic intrusion of souls into the natural evolutionary process, God intentionally designed a special human body patterned on the naturally evolved human life forms, which was to become the sole vehicle for soul incarnation on the earth. Souls were encouraged to incarnate into this special biological life form so that they might experience causality and thus become aware of the reason for their separation from God and aware of the mental state and mentally-directed physical activity that contributes to that ongoing separation.

Evolution had progressed from inanimate matter constructed from condensed raw spirit, to mind-directed animated matter called biological life to conscious-directed biological life by intelligent independent souls for the purpose of spiritual restoration.

**Physical evolution led to biological evolution
which is leading to spiritual evolution.**

Nonphysical nonbiological beings are using physical biological bodies to achieve spiritual rehabilitation and restoration to their rightful place in the family of God.

The Death of the Universe

But the day of the Lord will come like a thief. The heavens will disappear with a roar; the elements will be destroyed by fire, and the earth and everything in it will be laid bare. (2 Peter 3:10)

Peter is addressing the impatience of the followers of Jesus, who are expecting an imminent return of the Messiah after his resurrection and ascension. While many are anxiously awaiting this second coming and perhaps thinking that the Lord is a little slow in keeping his promise, Peter tells them that God is giving sinners the world over a chance to repent of their sins and accept the Lord before Jesus arrives because [as he believes] it will be too late after that event. God is patiently waiting so that as many people as possible will not perish with the arrival of Jesus. He also says that the second coming will be upon them like a thief in the night, and failure to prepare for it will mean that they are caught unawares and unprepared, so no one should think he can keep putting off the decision to accept and follow the Lord because the delay may mean their downfall. Peter engages in a little hyperbole at his point, claiming that the heavens will vanish in an instant and the earth will be burned up at the second coming. By "heavens", Peter does not mean the abode of God but the stars and other visible objects in the sky, the universe beyond the earth. Actually, the universe won't disappear with a roar in the sense that the universe will be destroyed by some major cataclysmic geologic event, but the earth will be destroyed by fire.

As the universe expands, the gravitational collapse of interstellar gases and dust to form stars will become a rarer and rarer phenomenon, and galaxies and stars will drift farther and farther apart. The black holes that occupy the centers of galaxies will continue to capture matter and energy and trap it in some unknown singularity where matter is crushed to infinite density and spacetime is distorted into a bizarre micro-region of infinite curvature. A place where nothing we know and understand exists or even makes sense. Galaxies will continue to race apart at an ever increasing rate of separation until light from one galaxy will not be able to reach a neighboring galaxy. The universe will ever so slowly turn dark and cold as the last old stars wink out, the remaining interstellar matter becomes too thin to clump into new stars, and temperatures approach absolute zero. Eventually, from any vantage point in the universe, the skies will be dark because objects will be falling away from each other faster than light can travel between them and gravity will become so weak that new stars will not form. The universe will go out with a whimper, not a roar.

> *Remember, as He gave, "The heavens and the earth will pass away - my word shall NOT pass away; but ye shall give ACCOUNT for every DEED done in the body!" Hence we find LAW. Law is love, yes - as love is law; yet God is not mocked, whatsoever ye sow, that must ye reap. [Edgar Cayce reading 69-4]*

The earth will indeed be laid bare and destroyed by fire but perhaps in a different manner than that anticipated by Peter. Within three billion years or less the sun will have grown larger and brighter, the earth will have become hotter, and all water and ice will have boiled away, leaving an uninhabitable planet. By about five billion years, after burning all of its hydrogen, the sun will undergo a period of greater expansion to form a red giant star that will set the stage for a collapse and subsequent re-expansion of its core. Its corona will expand to envelop the closer planets and possibly the earth, which will be scorched by fire. The death throes of the sun will follow the death of the earth as the sun goes through cyclical episodes of expansion and collapse. Because our sun doesn't contain enough mass to explode as a supernova, it will eventually exhaust itself in a final slow collapse into a white dwarf star. The surface of the earth will be scorched and burned during the periods of expanding corona but the atomic elements that make up the earth won't be destroyed by fire because atmospheric oxygen will have been stripped from the earth and the conditions needed to sustain elemental burning will not be present (Villaver and Livio, 2007).

Mathematical Descriptions of the Universe

Why is mathematics a language that describes forces and motion so well? Did God use mathematics to design the structure of the universe, to prepare a set of operational blueprints, to statistically predict its evolution from concentrated energy to matter particles to atoms to stars and galaxies? Did he tweak the initial conditions and parameters to increase the probability that an appropriate planet would be formed out of the chaos of stellar birth and death? Did he really design the helical shape and structure of DNA and seed the earth with cellular life so that it would evolve into a suitable habitat for each soul to face the repercussions of its own inappropriate activity? Is God quantifiable? Is mathematics only the language of physical laws, or is mathematics also the language of spiritual laws? Is it possible to mathematically describe and model the nature and activity of spirit or predict the movement and interaction of spiritual bodies? Is there really a solid boundary and clearly identifiable separation between the spiritual realm and the physical realm, or is it just an imagined boundary that fades away and dissolves as it is approached more and more closely by scientific instruments and the human capacity to reason and imagine? If the physical universe is an expression of God, a crystallization or condensation of spirit into less energetic matter and energy, then aren't all current mathematical descriptions of the behavior of matter and energy in the universe also descriptions of spirit that are valid for low-energy conditions? We may never know if God mathematically designed the physical universe, but we do know that the universe can be described by mathematics.

> *Train the entity in higher mathematics as will have to do with the electronics and dealing with the forces of the spheres. For the astronomy in the study of light, the study of the rays that are a part of each individual planet, each individual star, each individual asteroid are all a part of the forces in universal activities. [Edgar Cayce reading 4081-1]*

The equations of classical mechanics are applicable to the macro-scale world of large objects as opposed to the equations of quantum mechanics, which are applicable to atomic-scale objects. Non-relativistic classical mechanics can describe the motion of large objects moving at low velocities relative to the speed of light, but relativistic classical mechanics is needed to describe objects moving at a significant fraction of the speed of light. Isaac Newton's laws of motion (formulated in the late 1600s) that relate the movement of objects in response to an applied force and his law of gravitational attraction between material objects are laws of non-relativistic classical mechanics. These laws accurately predict the motion and mechanics of the world we are familiar with and the motion of planets and suns in their various orbits. An implied premise of classical mechanics is that there exists an absolute time and space independent of each other and any other physical properties of the universe. There is no time lag associated with the classical description of gravitational attraction; our planet feels the gravitational pull of the sun (93 million miles away) just as quickly as our sun feels the gravitational pull of the black hole at the center of the Milky Way galaxy (25,800 light years away). Classical mechanics assumes that the mass of atomic matter is a fixed quantity independent of speed; our cars don't get heavier as we drive faster. These ideas might seem to be intuitively correct but in fact they are false, only appearing true because we normally don't physically experience conditions where the limitations of classical mechanics become noticeable or obvious.

There has been an evolution and revolution in mathematical physics within the last 150 years. In the mid-1800s James Clerk Maxwell merged and reformulated the fractured understanding of electricity and magnetism into four equations that accounted for the new understanding that electricity and magnetism are two closely interwoven and related aspects of a single electromagnetic force. The new equations predicted the existence of electromagnetic waves and predicted that the waves would only propagate at one constant velocity, the speed of light, when traveling through a vacuum (space without the presence of matter). It was soon recognized that Maxwell's equations were applicable to large-scale phenomena (classical physics) such as the propagation of light waves through the universe but were not correct when they were applied to phenomena occurring at microscopic scales and atomic scales (quantum physics) such as the interaction between light waves and electrons. Forty years later, Albert Einstein would intuit a far deeper meaning to the nature of electromagnetic waves and would use the fact of their constant speed of propagation to postulate ideas that no one else had dared to imagine. In 1905, one of his several seminal papers that forever changed physics would explain the behavior of atoms in gases and liquids and three of his papers would use the properties of electromagnetic waves to reveal previously unimagined aspects of matter-energy interactions and relative motion.

Brownian motion was first described in 1827 by botanist Robert Brown after observing random fluctuations of microscopic pollen immersed in water. Eighty years later, Einstein thought about the behavior of those pollen grains and recognized in them empirical evidence of what some physicists (but nowhere near the majority) thought at the time, that matter is composed of atoms. He interpreted this type of behavior, the random motion of a microscopic object surrounded by a gas or liquid, as being caused by the transfer of vibrational energy during random collisions of the object with atoms or molecules experiencing thermal motion (Einstein, 1905a). This constant vibratory motion of colliding and recoiling atoms and molecules is detected and measured in gases, liquids, and solids as heat. In attempting to explain Brownian motion mathematically he devised a statistical method to estimate how far the pollen or other tiny object would move on average in a given interval of time (diffusion). Not only did he rightly interpret the motion as being caused by the impact of energetic water molecules, he also devised a relation that permitted the number of molecules in one gram molecular weight (the molecular weight of a molecule expressed in grams, also called a mole) of any medium to be calculated using the measured diffusion coefficient and viscosity of the medium and the size of the molecules comprising the medium. The existence of atoms and molecules was put on a solid footing and along with other theories and experiments prepared the way that led to quantum mechanics.

> *Vibration is movement. Movement is activity of a positive and negative force. [Edgar Cayce reading 281-29]*

The human body is ceaselessly bombarded by air molecules every second of every day, but it ignores these collisions when the ambient heat content, which determines the vibration rate of the air molecules, is similar to the heat content of the body at the surface of the skin. When the temperature of the air rises too far above the skin temperature, too much heat enters the body and the body can no longer maintain an optimum stable internal temperature and heat content. The body responds by opening skin pores and releasing saline water that can evaporate and reduce the skin temperature. When the temperature of the air decreases too far below the skin temperature, the body loses too much heat, and the internal temperature and heat content of the body declines below its optimum requirements. The body responds by closing skin pores and vibrating muscles to generate more internal heat. The thermal motion of air can bring comfort or stress to the body and invoke a mental command to adjust bodily processes. The body constantly experiences Brownian motion and has the ability to counter-act its effects when necessary.

One 1905 paper (Einstein, 1905b) helped usher in the field of quantum mechanics and a new set of physical laws (quantum physics) that bear little resemblance to the laws that are obeyed by macro-scale objects. It presented the first clear explanation of the photoelectric effect, the ejection of charged particles (like electrons) from the surface of a material (like a metal plate) when it absorbs high-energy electromagnetic radiation. The radiation energy must be large enough to eject an electron from the metallic surface, but experimental evidence showed that the energy of the ejected electron does not increase when the intensity of the radiation is increased. This behavior cannot be explained by the mathematics of classical physics where energy is presumed to exist in a continuous spectrum. Einstein, building on ideas previously postulated by Max Planck, explained the phenomenon by assuming that the energy in an electromagnetic wave is carried in discrete packets; that is, light energy is quantized. Light travels through space as a wave but exchanges energy with objects as an energy packet or particle (photon). All of the electromagnetic waves speeding across the universe or interacting with matter are vibrating at specific frequencies. When light strikes the surface of the metal, the energy of the wave, which is determined by its frequency, is transferred as a single photon to one electron, which is ejected from the surface of the metal if it is given enough energy. When a specific frequency of light has a greater intensity there are more photons available to interact with the surface electrons of the metal. This causes more electrons to be ejected, but all of the electrons have the same energy, and no electron is ejected with a different energy. Photons are now recognized as the force carriers of the electromagnetic field. They are the means by which the field interacts with particles. Einstein received the Nobel prize in 1921 for this revolutionary insight.

When electromagnetic radiation strikes the surface of a liquid or solid, it may not be energetic enough to dislodge an electron, but it may be absorbed into the material and cause an increase in the rate of vibration of an atom bound within the structure of the object. Like the energy imparted to the electron by the photoelectric effect, the energy of the sun's electromagnetic radiation is transferred across the surface of the human body by the absorption and emission of photons, much of it in the form of infrared radiation (heat radiation, a lower frequency than visible light), which warms the body as the energy is absorbed, and ultraviolet radiation (high-frequency waves), which is so energetic it can damage skin cells and cause skin cancer, especially in lighter pigmented skin.

Like light, all atomic-scale particles have both wave and particle properties. Light waves can behave like particles, and electrons can behave like waves. Quantum mechanics is the set of mathematical tools used to calculate and predict energy and matter interactions at the atomic scale. One of the fundamental relationships in quantum mechanics is the Heisenberg Uncertainty Principle, which states that certain related physical properties cannot both be measured to any predetermined accuracy. It is a direct consequence of the wave-particle duality. For example, the measured position and the momentum (mass times velocity) of an electron cannot both be known to an arbitrarily high accuracy, and the measured energy and time interval of measurement of a reaction cannot be known to an arbitrarily high accuracy. There will always be an uncertainty associated with the two paired measurements. In terms of understanding the creation and death of the universe, there are three important situations in cosmology where quantum mechanics is applicable and necessary to describe events; during the beginning of the creation of the universe when electrons, quarks, and atoms were being formed; in the interior of stars where thermonuclear reactions create heavy nuclei; and within the interior of black holes where large amounts of matter and energy are trapped and compressed by intense gravitational forces. Because quantum mechanics has been so successful in analyzing energy and matter interactions at atomic and subatomic scales, cosmologists have confidence that the equations of quantum mechanics are applicable to conditions found in the earliest minutes of the universe and can be used to predict the details of the early evolution of the universe. Quantum mechanics is the best current description of the behavior of the fundamental particles and forces of the universe and the interactions and reactions that occur among atomic and nuclear matter and energy.

Every electromagnetic wave has a specific vibrational frequency and wavelength (frequency is the inverse of wavelength), which determines the amount of energy held within the wave (the energy of an electromagnetic wave is directly proportional to the frequency of the wave). The universe is awash in electromagnetic waves vibrating at many different frequencies. Every quark has a mass that is a compact form of one half of the energy contained in the original electromagnetic wave that converted into the quark and its anti-quark sibling. Every electron has a mass equivalent of one half of the energy in the electromagnetic wave that converted into the electron and its antielectron sibling. These fundamental particles store vibrational energy as mass, the particle equivalent of wave energy.

Quarks are only found in nature in unstable pairs or stable triplets. When three quarks combine to form a proton or a neutron, they are held together by the strong force and move back and forth in random directions at nearly the speed of light, a motion that can be loosely interpreted as orbiting each other. This frantic motion within the core of every atom is a form of vibrational energy. Clusters of newly formed protons and neutrons clumped together in stable groups in the early universe to form the nuclei of the three lightest elements. Heavier proton and neutron clusters created in the interior of stars and during supernova explosions at the death of stars formed the remaining eighty-nine natural elements. Each cluster eventually attracted and acquired a menagerie of negatively charged electrons that surround the positively charged nucleus and balance the charge of the atom. These electrons are distributed within well-defined regions (orbitals) around the nuclei, each orbital having a different shape depending on the energy of the electrons, as determined by the Pauli Exclusion Principle, which disallows any two electrons with identical quantum properties from occupying the same energy level. This orbital motion of every electron in every atom has an associated vibrational energy or energy of periodic motion.

> *... all force in nature, all matter, is a form of vibration - and the vibratory force determines as to what its nature is ... as is seen in all matter in the earth's plane. [Edgar Cayce reading 900-448]*

Atoms of the 92 natural elements combine to form molecules and compounds that combine in various ways to form gases, liquids, solids, planets, dinosaurs, mice, human beings, chairs and tables, and everything else in the physical universe. Each physical object is a unique combination of atoms and the composite internal atomic vibrations are different for all objects. Just as there is a unique vibrational frequency and energy associated with the mass of each electron and quark and the orbital motion of electrons and quarks in atoms, so there is a unique vibrational frequency associated with the atoms and molecules in a gas, liquid, or solid. The atoms within every type of material in the universe, from the coldest man-made liquid helium to the natural ice sheets of Antarctica to the molten core of the mantle to the thermonuclear furnaces in the interior of stars, all vibrate as a unit according to the heat content of the object they form. Heat is the energy of atomic motion and is a measure of yet another vibrational component of matter. There are many other forms of motion energy on every scale observable across the vast expanse of the universe, including the spinning rotation of the earth about its axis, the orbital rotation of moons around planets and planets around stars, the regular pulsing of certain stars, the rotation of two binary stars about each other, and the rotation of stars within a galaxy around a central black hole. The spinning rotation of objects and rotation of objects about one another is a form of regular periodic motion that has a specific frequency and is a form of vibration with an associated energy of motion. The readings declare that everything in the universe is in motion and all motion is a form of vibration.

> *For, returning to the first principle, - as there are those forces that move one within another to bring harmony, as for light or color, or sound, or motion, all of these are but the variation of movement, vibration. [Edgar Cayce reading 2012-1]*

The development of quantum mechanics, the attendant discovery of the fundamental particles of nature, and the structure of nuclei and atoms verify that all atoms are a form of energy in motion. Each element, each isotope of an element, contains a specific assemblage of subatomic components, a unique combination of electrons and quarks that act together as a whole to form the element. Each electron and quark has the mass equivalent of a vibrating electromagnetic wave and every electron and quark is in constant orbital and vibrational motion within the structure of the atom. It is not that each element vibrates at a single pure frequency but that the vibration of its separate components creates a harmonious orchestral blend of all of their individual frequencies that is unique to that atom. As a collection of various elements, the body vibrates as a unit in some complicated pattern made up of the vibrations of its individual components. There are an estimated 37 trillion cells in a human body and an estimated 100 trillion atoms in each human cell, so there are many instruments in this orchestra. There are other modes of vibration associated with the human body. The readings largely agree with Hindu teachings on the presence of spiritual centers of activity in the physical body that are the connections between the physical body and soul, and they contain considerable information on the vibrational levels and basic impulses of the body's seven spiritual chakras and associated endocrine glands. The 281-series of readings supply a considerable body of material on prayer, meditation, and the endocrine glands and their interconnectedness and effect on the human body and the soul.

> *The body - EACH body - made up of vibratory force and of cellular conditions within the physical forces, that are as small worlds or universes within themselves, and the blood stream [is] that which correlates the ability of each of these atomic forces to work in unison throughout the system. [Edgar Cayce reading 121-1]*

The readings suggest that the lowest vibrational levels of the human body correlate with the fundamental physical response needed for survival of the body. The affiliated endocrine glands are the gonads, which are responsible for physical reproduction. This is the focal point of the self-centered biological response that ensures species propagation and can exist independent of any higher spiritual consciousness within the individual. At the other extreme, an individual that has gained a highly attuned spiritual consciousness operates at a higher vibrational level. This level is associated with the physical pituitary gland, which receives spiritual energy that has been concentrated in the pineal gland and distributes it through the body as it is activated by the soul's spiritual awareness of its oneness with God. A human being can be as close to an animal or as close to God as he lets his desires lead him. The color spectrum is an ordered list of frequencies or vibrations that coordinates with each chakra or endocrine gland and ranges from the lower frequency red color (gonads) to the higher frequency violet color (pituitary). Looking at this color spectrum in reverse order is suggestive of the formation of the universe and the reduction of the higher-frequency, more spiritual vibrations associated with its spiritual origin to the lower frequency more material vibrations of the mature universe of atomic matter (Todeschi, 2014).

Quantum electrodynamics (QED), a quantized version of Maxwell's classical equations of electromagnetism that successfully allowed the equations of electrodynamics to be applied to interactions between photons and charged particles on the atomic scale, was developed several decades after Maxwell's original work (Feynman and Zee, 2014). Like all quantum mechanical calculations, it is based on probability amplitudes (wave functions) because nothing in the quantum world is definite or exact. An example of a quantum mechanical interaction is the creation of an electron-antielectron pair from a photon and the movement of the created particles to some particular location in space. A wave function can be written that describes the possible location of the photon and the location and velocity of the particles after the interaction occurs. The probability that the particle will actually be at a specific location or have that velocity is calculated by multiplying the wave function by itself. A process can involve one or more interactions. When the process has been defined, how does one determine the probability that it will actually proceed as visualized? The probability that a given process will occur to an expected conclusion is determined by multiplying the probability wave functions for each interaction in the process that leads from the initial condition to the end of the process.

But this is only the beginning because there may be many ways, different sequences of interactions, which lead to the same result. Each of these independent ways is a unique potential history in the suite of photon and particle interactions that can arrive at the same result. The overall total and complete probability amplitude that predicts how the process occurs is determined by summing up the probability amplitudes of all of the potential histories that can lead to that outcome. The wave function of each possible step in each possible history extends throughout the universe, so each step in potential history can theoretically cause a photon or particle to end up anywhere in the universe, but there is a much greater probability that it will complete within the local vicinity where it began. Reality in the quantum world is best described by using a series of probability waves that incorporate all possible ways that an outcome can be achieved to predict the probability that a specific outcome will occur. Reality in the quantum world is not an ordered sequence of events that leads from one causal event in the past along a prescribed and definite path to one outcome in the future. Quantum mechanics does not predict the future from events that occurred in the past; it only determines the mathematical probability that a certain future will occur. The interaction of matter and energy at atomic scales is not deterministic and does not bear any resemblance to our conscious perception of the physical world or our mental concept of an expected logical behavior of matter.

The human body as a whole is not a quantum object but a classical object, a large collection of quantum objects bound together in a particular form. Its behavior and motion in the world is not described by quantum mechanics. What about the soul? It is composed of spirit, which according to the readings is a spiritual precursor of electromagnetic radiation that does obey quantum laws. Is the present state or condition of our soul in some way a sum of all possible histories that do not individually determine our future, but that in their totality are the equivalent of a quantum mechanical wave function that contains our probable movement along one spiritual path in the absence of any external interference? Will the next decision we make, the willful impetus that allows the mind to shape spirit, alter our present spiritual trajectory and set up a new probability amplitude of potential spiritual paths, some more likely and some less likely to be fulfilled?

Einstein shook the scientific world in 1905 when he published the special theory of relativity and forever destroyed the previous assumptions of absolute space and absolute time (Einstein, 1905c). The theory is special because it only describes physics in the special case when gravitational fields are absent and there is no relative acceleration between observers. The mathematics of the theory was formulated by applying two postulates to the situation of relative motion of observers in the universe. The study was constrained to the equations of motion that relate two different observers in frames of reference that move at a constant speed with respect to one another (for example, a person riding in a train and a person watching the train pass by). The first postulate is that the laws of physics are the same for everyone no matter where they are or when they are in the universe. Every part of spacetime exists on the same footing. No physical law singles out one location in space over another location or one moment in time over another moment. No location or time in the universe is more real or more important than any other location or time (Greene, 2005). Relativity declares that every moment is as real as any other, and that it is impossible to apply any physical measurement to identify any particular coordinate system as inherently stationary or uniformly moving. The second postulate is that the speed of light in a vacuum is constant for any observer, no matter their relative uniform motion with respect to each other or to the source of the light.

It is interesting to speculate that the property of the universe that makes the physical laws equally applicable to any material body anywhere in the universe might be in some way be a physical analog to the spiritual precept that God is no respecter of persons. Everyone is the same before God and has the same opportunity to approach him whether rich or poor, in a position of power or a homeless person wandering the streets. There is absolute equality of souls before the laws of God because every soul is equally a portion of God. There is absolute equality of material objects before the laws of physics because the laws of physics are identical at every location in spacetime.

Einstein's analysis showed that the previous concept of space and time as absolute and fixed characteristics and quantities of the universe was incorrect. The concepts of absolute space and time were never proven scientific ideas, but were accepted as unresolved issues because they were implied in the mathematics that accurately described interaction at a distance in classical physics. Relativity tells us that two observers at rest with respect to each other will measure the same length of objects and the same duration of time between events in their common environment, but if the observers are moving at a constant speed relative to each other, they will measure a different length in the direction of motion and a different time interval between events in their shared environment. They are forced to discard any preconceived assumptions about space and time having absolute and unique properties in the universe.

The entity will EASILY understand Einstein's theory. Few would! [Edgar Cayce reading 256-1]

The special theory of relativity tells us that our concept of space and time depend on the measurement process and the exchange of information via electromagnetic energy traveling at the speed of light. The observed and measured differences of space and time become larger as the relative speed between the observers increases. Space and time are not independent or fundamental properties of the universe but are two interwoven aspects of the spacetime continuum that forms the true foundational structure of the universe. A measurement of an interval of space necessarily contains a time interval component, and a measurement of time duration necessarily contains a space interval component. But two different observers in constant relative motion will measure the same spacetime interval. Spacetime, unified space and time, is the fundamental physical property that comprises the fabric of the universe.

> *When the heavens and the earth came into being, this meant the universe as the inhabitants of the earth know same; yet there are many suns in the universe, - those even about which our sun, our earth, revolve; and all are moving toward some place, - yet space and time appear to be incomplete. Then time and space are but one. [Edgar Cayce reading 5757-1]*

> *For each phase of time, each phase of space, is dependent as one atom upon another. [Edgar Cayce reading 3161-1]*

Special relativity determines what regions of the universe we can visit or interact with and what events in the universe we can see and in what sequence we will observe them. Every event, or action, has an associated light cone, and only regions of space and time that lie within that light cone can become part of our future. The speed of light is the speed of causality, and because it is finite, we are necessarily shut out of many regions of space and many eras of time. The physical effect of the finite speed of light is constantly with us, but our conscious minds do not typically perceive this property of the natural world. The effects of relativity are largely undetected by the physical body because it is only noticeable at speeds that are much greater that those achieved by macro-scale objects that are commonly encountered in the everyday experiences of humankind.

Reading 900-24 requests information on the law of relativity, but the answer has little to do with the special theory of relativity. It and many other readings discuss relativity in the context of the interrelatedness of objects in the cosmos and relativity of forces, which seems be a general expression that indicates that different parts of creation all have their relative aspects and activities. Each aspect of creation exists in relative condition to the others, a relativity of position and force that connects and ties the physical world together. The same relative force also applies to the soul in relation to God and man, the state of activity of the soul as it seeks to determine its relationship to the Creative Force. Another reading (4121-3) connects this relativity of force to the spiritual law like attracts like, whereby the mental attitude and activities of one individual create conditions that draw other like-minded individuals to them or within a common sphere of influence. None of the readings that mention relativity do so in the sense of the physical law of special relativity unless it is in the vague and broad context that the laws of the universe are equally applicable everywhere.

Another consequence of the special theory of relativity is that the inertial mass of an object (its resistance to a change in speed) increases as its speed increases. This insight led Einstein to conclude that matter and energy are equivalent and interchangeable and led to the publication of yet another paper (Einstein, 1905d). The most well-known and famous equation gleaned from Einstein's analysis is $E = mc^2$, which expresses the fact that the energy equivalent of the mass contained in every subatomic particle, nucleus, or atom is the product of its mass times the speed of light squared. The square of the speed of light is an enormous number, and the amount of energy contained in a single atom is huge. In this equation, mass is not a fixed value but is a function of the relative speed of the object to the observer. No material object can be accelerated to a speed greater than the speed of light because an infinite amount of energy is required to bring the speed of any object up to the speed of light. The conversion of energy into matter and matter into energy occurs constantly in the universe; for example, during electron-antielectron pair production from electromagnetic waves and during electron-antielectron annihilation to create electromagnetic energy. The conversion of matter into energy is quite noticeable when it is carried out in thermonuclear furnaces in the interior of suns where high pressure and temperature force protons together to form subatomic nuclei. The total mass of the created nuclei is less than the sum of the individual masses of its component protons, and that extra mass is converted to energy that radiates away from the sun.

It took Einstein another ten years to expand his special theory of relativity to include the presence of gravitational fields and accelerating reference frames. This general theory of relativity (Einstein, 1915) began with a realization that the attractive force of gravity is indistinguishable from acceleration in free space in the absence of gravity. The laws of physics to an observer in a closed room undergoing acceleration through interstellar space with no way to observe the outside world cannot be distinguished from the laws of physics when the room is in the presence of a gravitational field. This is tantamount to stating that inertial mass—the resistance of an object to a change in motion—and gravitational mass—the mass that is involved in gravitational attraction between objects—are identical. Using this idea, Einstein developed the general theory of relativity, which embodied a radically different concept of the cause of gravitation. General relativity explains how the distribution, density, and motion of energy and matter in the universe changes the geometry of spacetime and that what we perceive as gravity is actually an effect imparted to matter by distorted or curved spacetime, not by some intrinsic property of matter. Regions of higher gravitational force are associated with larger mass and energy concentrations. Matter causes spacetime to curve and curved spacetime causes matter to accelerate. Massless electromagnetic waves are not immune to this effect. Electromagnetic waves experience gravity just like matter because electromagnetic waves travel in the spacetime continuum and their direction of travel is influenced by the curvature of space-time. General relativity is the current best description of the large-scale structure and motion of the universe.

Another important insight from general relativity is that gravity, or curved spacetime, alters time. Clocks run slower in regions of greater gravitational force, or more curved spacetime. For the age of the universe to have meaning, time measurements in one part of the universe must be able to be synchronized with other time measurements at far-flung ends of the universe. Otherwise it would not be possible to assign an age to the universe as a whole, and instead, different volumes of the universe would have different ages because different mass and energy concentrations would affect clocks differently. This turns out to not be a problem. The early universe was characterized by extreme homogeneity, a nearly uniform distribution of matter and energy as shown by the fact that the cosmic microwave background radiation is uniform to a few parts in one hundred thousand parts. This uniformity meant that time could be synchronized throughout the early universe and time, which is associated with various locations on the spacetime fabric, does not get out of synch as spacetime expands. The synchronous time of the spacetime fabric has meaning at astronomically large scales and can be measured. The age of the universe can be determined, and one age applies to the entire universe. This does not mean that there is absolute time, only that the duration of the process of space expansion can be measured. The lapsed duration of time from the beginning of the expansion of the universe is not the same as time intervals measured by different observers moving through space relative to each other as given by special relativity.

There are several references to the attractive force of gravity in the readings. Almost all are in readings concerned with an attempt to develop a perpetual motion machine that would violate the physical law of energy conservation to generate free unlimited energy. In 1915, the mathematician Emmy Noether proved that symmetry in any physical theory is associated with a conserved quantity and that if there is time symmetry, the conserved quantity is energy. This has been verified as correct by many experiments. The premise of the perpetual motion motor design was that because everything is a force or form of energy and there is no true vacuum (no such a thing as nothing), a mechanical device could be devised to control the relative forces in water and air to extract energy (195-51 Reports). Like all perpetual motion machines, it violated the conservation of energy and failed to live up to expectations and generate abundant free energy out of water, air, or the vacuum of space. The descriptions of gravity given in these readings are convoluted and confusing in relation to current gravitational theory and cannot be correlated with the current scientific understanding of gravity. One almost wonders if the source of the readings was playing with the perpetual motion idealists.

Reading 2390-7 is more intriguing. It encourages an individual to hold fast in purpose and remain faithful to God and tells the individual that because Jesus worked in harmony with the spiritual and mental laws of God, he became master of the physical and spiritual laws. Jesus was able to work through the spiritual realm to affect change in the physical realm and thereby to accomplish miracles, feats that appeared to contradict the laws of nature. The reading states that Jesus was able to overcome the law of gravity, likely referring to the miracle of walking on water. The reading also states that the individual could accomplish the same feat as a result of the power that comes from holding true to the purposes of God. Peter was encouraged to emulate Jesus by walking on water, but sank in the water when fear and doubt crept into his consciousness (Matthew 14:22– 33). His conscious mind could not accept what his eyes were seeing.

The mathematics of general relativity has many similarities to the other equations of classical physics but is far more complex and quite difficult to solve for real-world distributions of matter and energy in the universe. Solutions have been found for certain special cases and approximations. Some physical phenomena described by the mathematics of classical physics, such as Maxwell's equations of electromagnetism, can be characterized in forms similar to the more complex mathematics of general relativity but that is usually not necessary for everyday applications at a macro scale. The law of gravitational attraction discovered by Sir Isaac Newton in the seventeenth century is a special case of the equation of general relativity when gravitational attraction is weak and the velocity of the gravitating objects is small. General relativity is applicable to large-scale phenomenon, events that take place over distances on the scale of planets, solar systems, galaxies, and galactic groups. It describes how the path of a light wave traveling at the speed of light can be bent as the wave passes near a large mass such as a star or galaxy, a process called gravitational lensing which has been observed in many high-resolution images from the Hubble Space Telescope. General relativity gave cosmologists the tools that allowed them to explain and predict observed events in the far reaches of universe including the behavior, influence and characteristics of supermassive black holes.

Human bodies certainly feel the effect of gravity as anyone who has fallen will verify. We have been taught to think of gravity as an attractive force that pulls us down toward the center of the earth but that concept doesn't fully grasp the essence of gravity and its effect on the human body. Gravity is better thought of within the context of spacetime as a tidal effect, an effect that occurs when a large massive object creates a curvature in the spacetime fabric, a gradient (a change of one quantity with respect to another quantity) of gravitational attraction in spacetime. The force of gravity is not constant throughout the body but seems that way because the body is rigid with respect to the weaker gravitational force it feels while standing on the earth. The force of gravity is actually stronger at the feet of a standing person than at the head. The force of gravity on each side of a body standing on the earth is not vertically downward but is downward and slightly inward toward the centerline of the body. We don't notice this tidal effect because gravity associated with curvature of spacetime near the earth isn't strong enough to change the rigid nature of the human body. But put that same body into the enormous gravitational field of a black hole and the story is different. As the body is pulled toward the center of the black hole, it elongates from head to toe and shrinks from side to side. The rigid mass of the body begins to stretch into a long thin spaghetti shape as it falls deeper into the black hole. The curvature of spacetime leads to some interesting non-intuitive effects.

Attempts to directly combine the mathematics of general relativity and quantum mechanics have not been successful hence the need for a completely new approach. Superstring theory sits at the cutting edge of theoretical physics. It is a mathematical construct that attempts to derive the properties of the known subatomic particles and force carriers of the universe from the vibrational characteristics of postulated elementary quanta of energy having the character of minute open or closed string segments. It is also an attempt to formulate a single theory that explains physical phenomenon at two extreme scales, the macroscopic scale where general relativity and gravitation rule, and the microscopic scale where quantum mechanics and the Heisenberg Uncertainty Principle apply. A theory of this type is necessary for science to be able to calculate and predict the physics of the universe where matter and energy are compressed by enormous gravitational forces into minute regions of high temperatures and pressures. These unusual conditions existed in the extremely high-temperature and high-pressure environment during the earliest moments of the creation of the universe when the inflation field began expanding and the first massless particles were being produced. It would also allow theoretical calculations of conditions in the centers of black holes where captured matter and energy are compressed into miniscule regions of extremely high matter density and highly distorted spacetime. A successful mathematical formulation of superstring theory would be a major step toward the development of a theory of everything, a single theory that would include all known physical phenomena (Greene, 2010).

The premise of superstring theory is that the fundamental matter particles and the force-carrying particles in the Standard Model of Particle Physics and the force-carrying gravity particle (graviton) can all be described as energetic submicroscopic vibrating strings having different tensions and vibrational rates. To achieve this, the mathematics of superstring theory requires the extension of spacetime dimensionality from the usual four dimensions to ten (nine spacelike and one timelike dimension) in several approximate forms or eleven (ten spacelike and one timelike dimension) in a form that strives to combine the approximate forms. These higher spacelike dimensions are envisioned as being compacted and periodic. One characteristic of the compacted dimensions is that they have smaller elemental lengths than our familiar three space dimensions. The postulated dimensions are so small that they are not observable and cannot be detected by current scientific instruments, rendering superstring theory impossible to verify by direct experimental testing.

Long before superstring theory was formulated the readings had described a multidimensional realm that souls experience as states of consciousness that are important to soul evolution. The formulation of superstring theory in higher dimensions, of which the three spacelike and one timelike dimensions of our universe are a subset, is somewhat reminiscent of descriptions in the readings of educational dimensions encountered during soul growth whereby higher dimensionality planes of consciousness (some associated with physical planets in our solar system) are staging areas and training centers for souls arriving and leaving the earth. There does not seem to be any way to correlate the higher spatial dimensions of superstring theory with the educational dimensions of consciousness described in the readings (see The Solar System section). However, the emphasis of the readings on vibration and motion as the foundational basis of the physical universe and the nonphysical realm is fully consistent with the basic principle of superstring theory, which postulates that all subatomic particles and force carriers are expressions of submicroscopic strings vibrating in different modes and combining to form all physically observable energy and matter. Superstring theory also allows for the possibility that the compacted periodic extra dimensions can occur in different arrangements that potentially lead to a separate and unique universes having completely different physical laws. It shares this predictive aspect with inflation theory which was formulated to explain the early behavior of the universe, even though these theories have nothing in common.

Creation Or Evolution

When it is considered that "a thousand years is as but a day and a day as but a thousand years in the sight of the Lord," (2 Peter 3:8) then it may be comprehended that this was colored by the writer in his desire to express to the people the power of the living God - rather than a statement of six days as man comprehends days in the present. Not that it was an impossibility - but rather that men should be impressed by the omnipotence of that they were called on to worship as God. [Edgar Cayce reading 262-57 paraphrased]

Using the Bible to justify slavery and racism may have been the most heinous abuse of scripture in America, but that is not the only way the Bible has been misused. Some church authorities today are so intent on the Bible being the religious theory of everything that they have concocted an entire "science" around its passages. Creationism or Creation Science, a paradoxical term designed to give legitimacy to one particular literal interpretation of the Bible, is the religious doctrine that presents the biblical story of the creation of the universe and life on the earth as scientific gospel and a form of truth that does not need to be researched or verified by scientists or nonbelievers. The biblical creation story was written billions of years after the formation of our earth by writers who never had access to modern scientific studies of galaxy and star formation, and who did not have the scientific training or equipment needed to accurately measure or understand the physical processes involved. This does not matter at all because it is so easy to fall back on the belief that the Bible is the Word of God, and therefore every statement in it is infallible. There is no leeway to allow for the fact that God used fallible people who did their best to present difficult concepts and sometimes expressed those concepts in symbolic form. The universe and the earth it contains were not assembled in a seven-day burst of creative activity out of a void 6,000 years ago to house two humans and their descendants or for the enjoyment of humankind. The earth is a typical by-product of the formation of stars, a process that has been active for about 12.2 billion years and may have produced millions of planets suitable for biological life. The earth was chosen from all of these contenders to be a temporary dwelling place of souls so that through expression and activity in human bodies they would have the opportunity to re-attune their minds to the Mind of God.

> *It would be a sign of great simplicity to think that the world was created in six days, or indeed at all in time; because all time is only the space of days and nights ... so that one must confess that time is a thing posterior to the world. Therefore it would be correctly said that the world was not created in time, but that time had its existence in consequence of the world. - Philo Judaeus, also called Philo of Alexandria (c 20 BC – c 50 AD)*

Creationists long ago abandoned the concept of the geocentric universe, which imagines the earth to be the center of the universe and all other celestial objects to revolve around the earth, but only after a concerted effort to ram it down the throats of the general populace through terror tactics such as torture, imprisonment, and excommunication. The concept is based on several Bible passages that describe the earth as stable and immovable (1 Chronicles 16:30, Psalm 93:1, Psalm 96:10). Nicolaus Copernicus amassed enough observational evidence to prove that geocentricism was false. He published his results in 1543 after several years of worrying about the consequences of publishing a heliocentric theory of the universe in contradiction to the doctrine of the Catholic Church. Apparently Copernicus was not aware that some early Greek schools of thought had already proposed that the earth revolved around the sun. In 1632 Galileo published his astronomical observations that confirmed and defended the heliocentric theory. His work was quite popular, which probably brought it to the attention of the Roman Inquisition. Galileo was tried by the Inquisition in 1633 and was found guilty of heresy, the crime of thinking outside the church box, and was sentenced to house arrest, where he stayed until he died in 1642. Despite the best efforts of the church, scientific observation and rigorous data analysis finally prevailed over religious dogma. Apparently that debacle did not dissuade the creationist cause.

Readers of the Bible have always been selective about which verses they will accept literally, which they will accept symbolically, which they think reflect cultural values of the day, or even which are, God forbid, in error. Why is it so easy for people to read the Book of Genesis and quietly accept a literal seven-day elapsed time period for the creation of the universe, including the formation of galaxies, suns, planets, and the earth's plant and animal life, including man, in the face of a massive body of evidence to the contrary? The processes of stellar and planetary evolution within the expanding spacetime framework of the universe are well known today. The age of the universe has been estimated at about 13.8 billion years, and the age of the earth is about 4.5 billion years, and the age of fossil evidence of earth life ranges from millions to billions of years. The inclusion of the biblical verse that states that a day is as one thousand years to the Lord (2 Peter 3:8) to bolster the literal interpretation of a seven-day creation does not make it more palatable in light of current facts and theories on the evolution of the cosmos.

It is ironic that the same people who quickly jump to the defense of a seven-day creation because it is written in the book of Genesis usually dance and tiptoe around the words of Jesus when he clearly explains to his disciples the connection between John the Baptist and the Old Testament prophet Elijah. At some point early in his ministry Jesus is approached by disciples of John the Baptist, who has been imprisoned by King Herod. John, who has spent his preaching career predicting and foretelling that the Messiah will come to set all things straight, has sent some of his followers to inquire about Jesus so as to verify that Jesus is the one who is to come. Like many Jews, John is apparently also looking for a strong leader who will deliver the Jews from the bondage of Roman occupation. Jesus basically tells them that his works and his words speak for themselves which lets them know that his mission is rooted in love and peace, not politics. Shortly thereafter, he speaks to a gathering crowd and extolls the virtues of John the Baptist. He ends his comments by stating emphatically that John was Elijah, who was predicted in the Jewish scriptures to reappear on the earth as a forerunner of the Lord (Malachi 4:5 and Matthew 11:7–15). To emphasize his statement he tells his audience, "He that has ears to hear, let him hear." In other words, he is telling them not to misconstrue or deny what he is saying and to just listen and accept. Christianity doesn't really want to accept these words of Jesus so Christians try to modify the words by stating that John the Baptist only exemplified the spirit of Elijah or that he just acted in a manner that one would expect Elijah to act.

After Herod murdered John, he began to hear of the activities of Jesus and wondered if Jesus was John who had risen from the dead. Bringing up the subject again while ministering in Caesarea, Jesus asks his disciples what the people believe about him. This is the famous passage where Peter announces that Jesus is the Christ and makes a fundamental mental leap about his true nature (Matthew 16:16), but it also mentions that people thought Jesus may have been an incarnation of John the Baptist, Elijah, or Jeremiah, indicating many people believed in the possibility that an individual could return after death to another life. Sometime later, Jesus takes Peter and the brothers James and John to a high place where he is transfigured, transformed into a being with a brightness that rivals the sun, and talks with Moses and Elijah. The vision of Elijah stimulates memories of the previous discussions about John the Baptist, so his disciples ask Jesus about the words of the scribes, those who copied religious documents in the days before the printing press was invented, that predicted Elijah would come first. Jesus agrees that Elijah must come again to "restore all things" but follows with the statement that "Elijah already came, and they did not recognize him, but did to him whatever they wished" (Matthew 17:10–13), and his disciples understand that Jesus is referring to John the Baptist. With this statement Jesus reinforces his previous statement about Elijah, and the disciples understand that Elijah reappeared on the earth as John the Baptist and was killed at Herod's command.

Elijah truly did come to set things straight before Jesus began his mission, as was foretold in scripture, but they did not recognize him in the form of John the Baptist. When Jesus states something so clearly and reinforces his statement when it is obvious his disciples were having difficulty accepting or understanding his words, what right do we have to try to twist his meaning back into a form with which we might be more comfortable or which might be more compatible with current doctrine? The soul of Elijah did indeed come again to quicken the body of John the Baptist and give the forerunner his spiritual strength. It did not incarnate into or recreate the same body as Elijah, but the soul that animated John the Baptist was the same soul that incarnated as the prophet Elijah who lived and served God during the ninth century BC. Interpretations of biblical passages are influenced by many factors, including previous religious training and secular education, and often simply reflect a desire to read certain preconceived ideas into or out of the text. It takes effort and an open-minded attitude to discern the spirit and scripture correctly.

During the first two decades of the twentieth century, many religious authorities and scientists were finding common ground by being in general agreement about evolution and natural selection as it applies to the natural world. The artificial separation of religion and science over the theory of evolution was not as pronounced as it is today. Creationism has its origins in twelve tracts, called the Funda-mentals, published by A.C. Dixon from 1910 to 1915, which promoted religious conservatism, insisted on scriptural literalism, and rejected modern biblical interpretation. These tracts in turn have their roots in mid-nineteenth century Millennialism and similar isms that embraced such ideas as a 6,000-year-old earth, the need to prepare for an imminent apocalypse, and the brand new concept of rapture. The Fundamentals were basically a reaction to a perceived and feared end to the proper order of society by a flood of immigrants, a feared dilution of the predominately white Anglo-American majority and the increasing divide between rural poorly educated and urban well-educated segments of society as the nation transitioned from an agrarian to an industrial society (Principe, 2006). Some of these concerns still sound familiar today and are rallying points for certain portions of American society, especially members of conservative religious organizations and right-wing conservative political circles.

The age of the universe and the age of the earth that is accepted by most fundamental Christians has changed now and again depending on whose calculation is believed, but all of the proposed ages basically are variations on the calculation published by Archbishop James Ussher in 1650 using genealogical data extracted from Bible verses. His calculation definitively showed that the first day of the six-day biblical creation of the earth was October 23, 4,004 BC according to the Julian calendar used at that time. The slightly more accurate Gregorian calendar, used by most countries today, corrects an error in the length of the year used in the Julian calendar and slightly changes the calculated date. Interestingly, the age of the earth as deduced by Jewish scholars over the past several centuries using the scriptures of the Jewish faith ranges from 40,000 years to 2.5 billion years. Of course, the Jewish scriptures are the same Old Testament passages used by Christian literalists. Today, Christian-supported providers of home-school curricula and university courses dream up textbooks that contain amusing paragraphs where the formation of coal is a result of the compaction of trees that were swept into valleys during a worldwide flood at the time of Noah without a shred of research or physical evidence to back up such statements (Bob Jones University, 1990). Such dishonesty is deemed acceptable when it is used to attack science and support theology.

Religious pseudoscientists continue to devise ways to compress the entire 4.5 billion-year lifespan of Planet Earth into 6,000 years despite the massive body of data that supports the more ancient age. They deduce their untested conclusions from a few obscure genealogical passages mentioned in the books of the Old Testament that list the years that various kings reigned and that Adam and his immediate descendants lived. They expend much effort to show that dinosaurs, which lived from 250 Mya to 65 Mya, coexisted with men. Christian groups make Christ-oriented films "meant to reach a vast number of people who have been misled into accepting the evolutionary theory and thereby have come to doubt the forthright statements of the Word of God concerning man's origin, salvation, and eternal destiny," as if pseudoscience is essential for the salvation of mankind (Taylor, 1973, Morris, 1980, and Cole and Godfrey, 1985).

From personal experience, Christian summer-camp counselors explain the presence of fossils in geologic strata to children as party favors sprinkled through rock so that God's children can enjoy the excitement of a treasure hunt or as Satan's strategy to confuse and mislead the faithful. It should not be acceptable to lie and deceive young people for the purpose of keeping them in the faith or out of fear that they might entertain the heretical thought that biblical science might be wrong. Surely, forcing the reality of a 13.8 billion-year-old universe and a 4.5 billion-year-old earth into a predetermined 6,000-year-old worldview can't be evil if it is done in the name of God. The subversion of truth through false science textbooks is easy to do, especially in the context of home schooling, and is a convenient method for churches to exert their spiritual authority as the keepers of the faith and the holy Word of God. Church authorities also attempt to discredit accepted scientific methods such as age-dating techniques that yield evidence about fossils that is counter to biblical literalism by claiming that the techniques are flawed without offering a shred of evidence. The readings claim that such analytical methods do exist and should be used to study the past and suggest lines of investigation that appear to be beyond the capability of current scientific techniques. The following reading not only encourages the inquirer to develop these scientific analytical abilities but indirectly infers that animals inhabited the earth for at least ten million years.

> *How many of those that usually open an egg that's been buried for five to ten million years can, by its analysis, tell you what it's composition is, or what the fowl or animal fed upon that laid it? This body can! Get the idea? If there were opened a sarcophagus in the form of an activity where the feet of such an animal or man or body were chemically analyzed, as related to the odors and their active influences upon the body itself, would it not be possible to indicate as to what had been the means or activities of such a body during that experience in the earth? and as to what were the principal food values? [Edgar Cayce reading 274-7]*

Scientific discoveries made by careful observations, formulation of theories, and repeated testing of the predictions made by those theories typically stand the test of time. But too often they are blindly refuted by Christians who choose to unquestioningly accept pseudoscientific interpretations of biblical passages from their church leadership as being the truth without making any attempt to assess the validity of those interpretations. It is easier and perhaps preferable for many to be enlightened by the leaders of their particular religion rather than to put in the due diligence work required to verify the facts, thereby risking the stigma of heresy from the church to which they declare allegiance and from which they seek spiritual sustenance. Admittedly, truth that is sought and approached from the physical conscious mind and its limited perspective is elusive, but that is why science relies on the scientific method, not on the pronouncements of individual scientists.

Creationism has taken a beating in the past few years, but the concept of intelligent design is being promoted by some pseudoscientists as its successor. It doesn't seem to be an attempt to disprove scientific theories of the origin of the universe and alter scientific facts, but is more of an effort to reinterpret the creation of the universe and life on earth from an accuracy and precision approach. It is based on the premise that God is the creator of the universe because only he is able to set the fundamental constants of the universe to the extremely high precision needed to make all matter and life possible. Of course, intelligent design excludes the possibility of evolution and natural selection as the driving force of biological life. It is based on the assumption that recognizable patterns and organization in the natural world are evidence for intelligent activity, specifically the activity of God, and so, to some extent, sidesteps the need to negate most secular science or set up a separate biblical "science" to compete with the theories of academic scientists, except any scientific studies that should confirm the process of evolutionary advancement of biological organisms.

The basic concept of intelligent design is that God suitably adjusted or fine-tuned certain fundamental physical constants of nature, such as the charge and mass of an electron and Planck's constant, during the creation of the universe so that it would be able to generally support life and specifically support human life. This idea seems to be homocentric; it puts the human body in the form of Homo sapiens, which was not created until about 13.8 billion years after the universe was set in motion, as the centerpiece around which the edifice of the universe was constructed. In other words, it is like God had the perfect body in mind or on the drawing board and then calculated the precise fundamental physical constants required to initiate the first burst of creative activity so that after 13.8 billion years the last piece of the puzzle would slip into place and the earth environment would support a human body. This idea assumes a purely deterministic universe in which every event from the first instant of creation is determined completely by previously occurring events. This creation concept completely ignores the probabilistic aspects of quantum mechanics that describes all physical activity at atomic and subatomic scales. Interactions within and among atoms are not deterministic but are indeterminate and probabilistic. The early universe, at least until the creation of atomic matter, was a quantum universe.

Intelligent design seems to hearken back to the geocentric verses heliocentric descriptions of the universe. The church believed that the earth, as the abode of humans, is obviously of such supreme importance that God must have created the universe around it, and the promoters of intelligent design believe that the human body is obviously of such supreme importance that God must have created the universe around it. It is interesting to consider that the specific human body that we occupy may not be necessary to fulfill God's plan for soul growth. Perhaps another life form that can be soul controlled and through which souls can interact with love and kindness and otherwise express their mental development and attitude would work equally well. It does not seem reasonable that external appearance or form should be the primary qualification for beneficial soul-directed activity and soul growth.

The equation of general relativity shows how spacetime curvature is related through the constant of gravitation and the speed of light to a particular distribution and motion of matter and energy. As of 2020, the most accurate measurements of the gravitational constant have yielded an accepted value of $(6.67408 \pm 0.00031) \times 10^{-11}$ m3 kg–1 s–2 (cubic meters per kilogram per second squared) in metric units. The uncertainty of the number is about three in ten thousand. Because gravitational attraction is so weak and there is no way to shield an experimental apparatus against unwanted gravitational attraction from the surrounding environment, it is difficult to accurately measure the gravitational constant (Xue and others, 2020).

In 1917, Einstein added a negative cosmological constant term to the original equation of general relativity to counter the effect of gravity and make the equations predict a static universe that would not collapse back on itself at some distant future time. This constant caused the equation to conform to the generally accepted scientific understanding that the universe is a stable system, neither expanding nor contracting. But more than a decade later, astronomical measurements by

Edwin Hubble confirmed that the universe was in fact expanding, and Einstein effectively discarded the constant by setting its value to zero, which allowed his equation to show an expanding universe. In a sense, scientists fell into the same trap as the Catholic Church several hundred years before, but instead of declaring that the earth must be stable, they had decided that because there were no measurements to the contrary, the universe must be stable. Another seventy years later, more accurate astronomical observations revealed that the rate of expansion of the universe is not constant but is accelerating, which indicated the need to revive the cosmological constant but with a positive value. The new constant was added to the equation and is now being reinterpreted as the quantum mechanical vacuum energy of space and is thought to be associated with dark energy (Mohr and others, 2016). Its value has not been accurately determined but is about $(1.1056 \pm 0.0227) \times 10^{-52}$ m-2 (per meter squared) in metric units.

Films designed to promote intelligent design lead the viewer through an ever-expanding series of zeros following a decimal point or mind-boggling powers of ten to show that the fundamental constants of nature must have been set by God with fantastic accuracy and precision or the universe as we known it could not have formed, and only God has the ability to create a universe that needed that much fine-tuning of its fundamental constants. The film creators not only know this is true but also know the accuracy to which God set these constants. He set the gravitational constant to an accuracy of one in sixty decimal places (or no stars could form) and the cosmological constant to an accuracy of one in one-hundred and twenty decimal places (or the universe would expand too rapidly or too slowly). Actually, the expansion or contraction of the universe depends on the sign (positive or negative) of the cosmological constant, not the accuracy of its value.

But it gets more amazing or bizarre, depending on your perspective. One film (Craig, unknown date) declares with certainty that when the universe was formed, if the mass and energy had not been evenly distributed to a precision of one part in ten to the tenth power to the one-hundred and twenty-third power (a mind-numbing meaningless number that is beyond imagination as relating to any known physical constant or property of nature), then there would have been no matter or energy. In reality, the distribution of mass in the early universe varied by about one in one hundred thousand parts, and without this variation the first stars would not have formed. This slightly uneven distribution was essential for the future creation and evolution of the celestial objects. The creators of the film fail to report how these amazingly accurate values relate to the fact that time is not physically meaningful until approximately 10^{-43} seconds after the creation of the universe, and the concept of space does not have physical meaning at intervals of distance smaller than about 10^{-35} meters and that even the smallest quantities currently being investigated by physicists cannot be measured to anything approaching an accuracy of sixty decimal places (Susskind, 2006). The predictive capability of the Standard Model of Particle Physics holds true at a high degree of accuracy without the need to know any of the properties of the matter particles and force particles and associated constants to anything near one in sixty decimal places.

It is impossible to imagine that God needed to set the fundamental constants of nature to a minimum accuracy of one in sixty decimal places, let alone the incredulous one part in ten to the tenth power to the one-hundred and twenty-third power. It is utterly amazing that the film authors are aware that these constants were set to this exceedingly high degree of accuracy when the scientific community that studies atomic and nuclear processes and the forces that create stars and galaxies are clueless as to the true and absolute values of these constants to anywhere near that accuracy. Perhaps scientists need to seek out the film makers to learn how to measure physical properties to this degree of accuracy so that they can greatly improve their scientific instruments and calculations. It is ironic that having failed in the past to compel people to believe in a geocentric universe by force of the harsh and unforgiving Inquisition and threat of excommunication and having failed to prove that God created the universe 6,000 years ago, some of today's religious authorities still feel the need to prove or support the fact of his existence and creative powers by using doctored and manipulated constants of nature and particle physics. Science seeks truth by observing, studying, measuring, and testing the physical properties of the universe without deception or deceit, but too many Christians apparently fear that the truth of the nature of the material universe will in some way harm religion and deceive the faithful. Only that truth that corresponds with their literal interpretation of scripture or special insight into the mind and actions of God will be allowed to stand.

The readings actually support the general supposition of creationism and intelligent design, the idea that the universe and man were created by God, but they do not try to force science to be subservient to the information they present or to literal interpretations of biblical passages. The readings are generally compatible with the Darwinian Theory of Evolution, except when it comes to the origin of the specific human body type that souls currently occupy and the relationship between soul-directed human beings and the natural world. Evolution is a generational process whereby species of living organisms adapt to their environment over time and genetically transmit the more favorable adaptations to their offspring. Species that don't adapt, or that are too inefficient relative to competitors, die out. This is typically a slow process as applied to large animals with few annual offspring. Abrupt changes in environmental conditions can endanger or eliminate a species that cannot adapt quickly enough or whose normal range of habitat is too limited to allow them to migrate to avoid or recover from a major environmental change. This process is called natural selection and was proposed by Charles Darwin in 1859 after years of extensive research on biological diversity and the fossil record. The phrase "survival of the fittest" is not a statement of the relative physical strength of competing animals but is a statement about the fitness of individual species to adapt to the natural environment in a way that gives them a reproductive advantage over competing species. Our present body, whose creation is poetically and archetypically described in the biblical story of Adam and Eve, was patterned on existing higher life forms but was created outside of the ongoing evolutionary process specifically to provide an appropriate vehicle for the use of the soul in an environment suitable to spiritual growth.

Once it was created and placed in the earth's environment, the new body type became subject to the physical laws of evolution, but soul-guided man is not completely subject to the same forces of nature that impose the law of natural selection on all other species (900-340). Man is animated by a soul that has the ability to influence and alter its environment in ways unimaginable to other higher life forms as needed to ensure the survival of the body or simply to make life more comfortable. Unlike plants and animals, the will of the soul acting on the mind and manifested through the body interferes with and interrupts its natural evolution. The survival of mankind does not depend on how well he competes with the activity and behavior of other animals but is predicated on the extent to which he is willing to learn about God and desires to manifest the spiritual force that his soul represents and that underpins all life on the earth (900-340). Man does not just react to his existing environment, but can actively change his environment and therefore his physical destiny. He is extremely capable of manipulating and damaging his environment to the point where it will no longer sustain the human body. The next few decades will determine how seriously human-caused climate change, the unintended disruption of the natural cycles of climate and weather, will impact man's ability to produce food, and whether loss of plant and animal biodiversity will irreparably damage the ecosystem that is essential for man's survival.

> *The needs of those in the North Country [are] not the same as those in the Torrid region. Hence development comes to meet the needs in the various conditions under which man is placed. He [is] only using those laws [evolution] that are ever and ever in existence in the [earth] plane. ... The theory [Darwinian Theory of Evolution] is, man evolved, or evolution, from first cause in creation, and brings forth to meet the*

needs of the man, the preparation for the needs of man has gone down many, many thousands and millions of years, as is known in this plane, for the needs of man in the hundreds and thousands of years to come. [Edgar Cayce reading 3744-5]

... the SOUL OF MAN is that making him above all animal, vegetable, mineral kingdom of the earth plane. Man DID NOT descend from the monkey, but man has evolved, resuscitation, you see, from time to time, time to time, here a little, there a little, line upon line and line and line upon line. [Edgar Cayce reading 3744-5]

The natural world is the setting in which biological life evolves, in which plants and animals adapt to changing environmental stresses. Our lives on the earth are not just a result of an evolutionary biological drive to propagate the human species. We are not purely biological beings with every infant only a product of the genetic sequence it inherited from its parents. We exist as souls temporarily using a body to interact in a material realm, and our souls are far more important than the bodies we inhabit. The strength of the soul greatly exceeds the strength of the physical body. We were not born to our parents as a matter of chance nor do we experience the triumphs and tragedies of life because we were in the right place at the right time or the wrong place at the wrong time. Personal, social, and environmental stresses on human bodies and soul minds are used to awaken the soul to its separation from God and offer it the opportunity to modify its behavior and return to its original state of harmonious union with the consciousness of God. How long this process will take depends on the soul, but God in his infinite patience has made available the environment needed for soul growth, and according to reading 3744-5, it will be available for thousands of years to come, but perhaps the readings should have included the caveat; if we don't destroy it first.

Soul Incarnation And The Rise Of Man

The first humanlike animal, or animallike human, and forerunner of man was a pre-Australopithecine species with new characteristics and behavior that distinguished it from chimpanzees (Figure 2). There is some fossil evidence, although limited, that by about 7 Mya a pre-Australopithecus species that was living in central Africa had a small brain and prominent brow ridges, was bipedal (it could walk upright on two legs), and had developed a non-honing denture arrangement that was different from its predecessors. The two canine teeth located in the upper jaw did not have an inner blade ridge that sharpened during slicing and chewing when these teeth ground against the corresponding teeth in the lower jaw. The first two characteristics are apelike, and the last two are humanlike. These species retained other apelike features, including a tilted face with a protruding jaw and low forehead. Three species have been identified from fossil evidence found in central and east Africa and were dated from about 7 Mya to about 4 Mya. The species that was most notable lived from about 5.6 Mya to 4.4 Mya and did not have a grasping toe, which indicates the beginning of a transition from tree dweller to tree-ground dweller to pure ground dweller. An important characteristic of the pre-Australopithecine species was a small 350 cc (cubic centimeter) brain similar in size to a modern chimpanzee (Larsen, 2019).

Human Ancestry	Time (mya)	Brain Size (cc)	Geologic Epochs Readings information (Time in mya)
Hominini tribe	10.0 to 7.0		***Miocene Epoch*** 2665-2 Soul activity?
pre-Australopithecines S. tchadensis	~7.0 to ~4.0 ~6.0	 350	
Ardi. kadabba	5.8 to 5.6		***Pliocene Epoch***
Ardi. ramidus	4.4		Souls experiencing earth in non-physical bodies?
Australopithecines	~4.0 to ~1.0		
A. anamensis	~4.0	430	
A. afarensis	3.6 to 3.0	445	
-A. aethiopithecus -A. boisei	~2.5 2.3 to 1.2	430 520	
-A. africanus -A. robustus	3.0 to 2.0 2.0 to 1.5	450 500	
-A. garbi	~2.5	450	Souls experimenting with mineral, plant and animal kingdoms?
Homo habilis	3.0 to 1.8	650	
Homo erectus	1.8 to 0.30	950	***Pleistocene Epoch***
Homo sapiens Early Archaic Late Archaic *-Neanderthal*	 0.35 to 0.13 0.35 to 0.03	 1250 1450	Souls incarnating into higher life forms?
-Irhoud 1 & 2 Early Modern	~0.315 0.20 to 0.01	1350 1500	 0.212 -Atlantis 0.106 -Amilius/Adam
Late Modern	~0.01	1500	0.012 -Atlantis sinks

Figure 2. Correlation of anthropological time, brain size, and readings information. (Anthropological data extracted from Larsen, 2014, and Emiliano and Osbjorn, 2013)

Fossil evidence from central Africa indicates that by about 4 Mya the pre-Australopithecine species had evolved into an Australopithecine species, of which Australopithecus farensis was the major species in central and east Africa by 3.6 Mya to 3.0 Mya. This new species had a somewhat larger brain at 430 cc but was unable to speak because it did not have a developed hyoid bone, the small bone in the neck that is necessary for speech in humans. Over the next two million years Australopithecine would divide into three different subspecies, one of which would lead toward the first humans. These three groups did not arise in sequential order but were more or less contemporaneous. Fossils of the three species were found in different regions; east Africa, central Africa, and south Africa. Their physical characteristics were similar and the relative length of their arms and legs were similar to modern humans. The brain size of the original species and three derivative species ranged from 410 cc to 430 cc. All of these Australopithecus species had disappeared from the earth by about 1 Mya.

Human ancestors identified as the genus Homo split away from the Australopithecines between about 3.0 Mya to 1.8 Mya depending on which of the two Australopithecus species is taken to precede the first Homo species. Fossils of Australopithecus garbi that appeared in east Africa about 2.5 Mya was found in a setting that suggested the use of tools (cut marks on bones) but no tools were found at the site. This species had a brain size of 450 cc and is suspected to be the predecessor (along with Australopithecus afarensis) of Homo habilis, the first truly humanlike species to evolve from Australopithecus. Homo habilis fossils, which were found in central and east Africa, indicate a flatter face with large brow ridges and bipedal locomotion on short legs. They were found with stone tools. The brain size had increased 45 percent from Australopithecus within the relatively short timeframe of about 1 million years. The successor to Homo habilis was Homo erectus, the species that exploded out of east Africa and in a short time could be found in central Asia, Java, and Europe. Asian fossils have been dated from about 1.8 Mya to 300,000 years ago and the European fossils from about 1.2 Mya to 400,000 years ago. This bipedal species had a larger brain size than its ancestors, had a 30 percent larger body size (attributed to successful meat hunting), had a modern arm-leg ratio and large brow ridges, and used tools, including a new tool; the hand axe. Fossils show a sagittal keel, a slight ridge extending front to back across the top of the skull. Homo erectus fossil sites show the first evidence of controlled use of fire and brain sizes are about 900 cc, indicating continued rapid growth in brain size possibly related to speech, tool making, and coordinated hunting in groups.

By about 300,000 years ago human ancestors had evolved into Homo sapiens, characterized by an increasing trend toward flatter faces; higher, more vertical foreheads; rounded and taller skulls, and projecting chins. Brow ridges were on their way to becoming much smaller. The first of these new species (Early Archaic Homo sapiens) lived from about 300,000 years ago to about 130,000 years ago and was characterized by a larger brain size (about 1,200 cc), smaller teeth, and increased cultural complexity. It retained the larger brow ridges of its predecessors. This species transitioned into the Late Archaic Homo sapiens that lived from about 130,000 years ago to 30,000 years ago. This group includes the Neanderthals that lived throughout Europe and Asia. They have many characteristics in common with modern humans including the use of symbols such as cave art and body ornaments. They started the practice of burying the dead and probably had the ability to speak, as is indicated by the presence of fossil hyoid bones in Neanderthal remains. Whether Neanderthals could actually use this anatomical structure to speak is still an open question.

Neanderthals had a larger brain size (1,500 cc) than their ancestors. Their short thick bodies and large nasal cavities are probably an adaptation to the dry cold climates associated with the extensive glaciation of northern Europe and Asia during their reign on the earth. The Neanderthals retained the larger brow ridges, protruding teeth, and lack of chin that characterized their ancestors. Fossils discovered during recent excavations at Jebel Irhoud in Morocco were difficult to categorize because they had both Neanderthal and modern human characteristics and are thought to be a transitional human form (Emiliano and Osbjorn, 2013). The average brain size determined from craniums discovered at the site was about 1,350 cc. The find was unusual because Neanderthal DNA is mixed with modern human DNA in European and Asian populations at about one to four percent, but Neanderthal DNA does not occur or occurs only in trace amounts in modern African populations.

The Neanderthals were very likely the predominant human species to occupy the continent of Atlantis during the early periods of the continent's history as described in the readings. Archeologists have not identified remains of an Atlantean culture although the readings claim that residual evidence of Atlantean culture is plainly visible in the ruins of ancient civilizations throughout North America, Central America, South America, the Nile valley of Egypt, and the Basque region of Spain. Geologists have not discovered evidence of a previous mid-Atlantic continent now buried beneath the Atlantic Ocean. These avenues of investigation are not pursued by mainstream scientists because historical references to Atlantis are scant (Plato, 2008) and stories about Atlantis are generally believed to be a myth. Securing funding for such studies would be nearly impossible and few scientists are willing to risk their reputations in pursuit of Atlantis. However, there are many readings that delve into Atlantean culture and geologic changes to the continent as it broke up and sank into the Atlantic Ocean over a period of several tens of thousands of years within the context of the detrimental effect of selfish incarnated souls on society and the earth.

The layman concept of the Neanderthal as a brutish, hairy, ugly, stupid, and somewhat apelike creature unable to speak is based on century-old speculations from the first few fossil discoveries. The Neanderthals coexisted with and were followed by Early Modern Homo sapiens, who lived from about 200,000 years ago to about 10,000 years ago, and Late Modern Homo sapiens, which are only about 10,000 years old. By

40,000 years ago, the Neanderthal and modern humans were coexisting throughout Europe and Asia. DNA evidence indicates that the modern Homo sapiens species did not replace the Neanderthals but that the Neanderthals were assimilated into the modern species. A recent study of a 40,000-year-old Early Modern Homo sapiens fossil from Romania surprisingly revealed that it shared six percent to nine percent Neanderthal DNA. Certain aspects of the chromosomal structure indicated that the individual had a Neanderthal ancestor as recently as four to six generations back (Fu and others, 2015). Although fossil evidence is scant, the use of advanced DNA analysis and new techniques to extract DNA from fossil bones indicates that modern humans also interbred with Denisovans, an extinct relative of the Neanderthals, whose remains have been found at the Denisova Cave in Siberia.

The evolutionary advancement of human ancestors from Homo habilis to Late Modern Homo sapiens was remarkably fast, especially the rapid (on an evolutionary time scale) increase in brain size. During the approximately four million years that the pre-Australopithecines and Australopithecines species inhabited the earth the average brain size of these species increased from about 350 cc to about 500 cc at an average rate of 34 cc per million years (Figure 3). From the rise of Homo habilis about 3 Mya to Early Archaic Homo sapiens, the brain size of the Homo genus increased from about 650 cc to about 1,250 cc at an average rate of 227 cc per million years. Extrapolating backward from 3 Mya at this same rate of brain-size development (dashed line in Figure 3) intersects the pre-Australopithecines and Australopithecines brain-size growth line at 4 Mya, perhaps marking the divergence of these groups. The brain-size increase from Early Archaic Homo sapiens to Early Modern Homo sapiens is even more remarkable, increasing from 1,250 cc to 1,500 cc within a span of about 150,000 years.

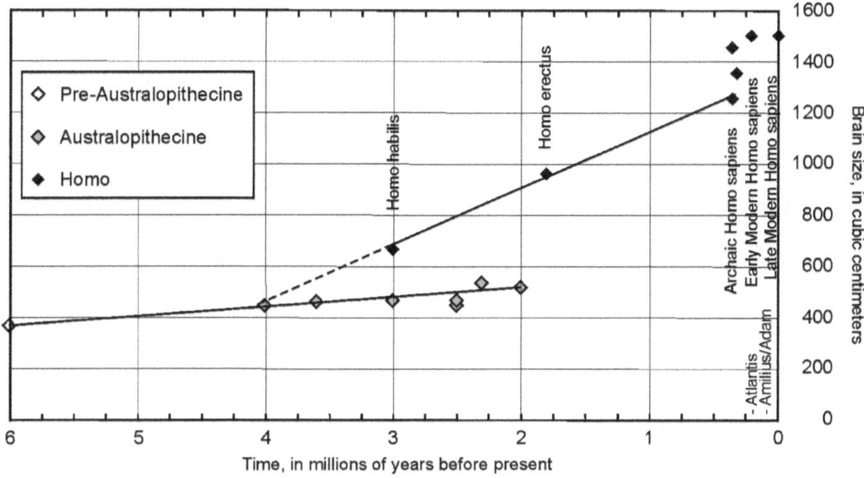

Figure 3. Temporal changes in brain size in pre-Australopithecine, Australopithecine, and Homo species.

Why is there such an abrupt divergence in man's evolutionary history as related to brain size, and why did one species set off on such a remarkable rate of brain-size evolution while a similar species dwindled and perished? In light of the readings, it is quite tempting to speculate that this divergence and abrupt increase in the rate of brain development was in some way related to an increasing interference of souls in the Homo genus human form that developed on the earth through evolutionary processes. Were souls intruding into ancestral human body forms that were similar to ours today before God was ready to create a better vehicle for soul incarnation? The chaotic situation was not conducive to the plan to use the earth as a realm of soul experience for the application of spiritual principles that would to lead to soul evolution. A proper body form and the restriction of soul incarnation into that form would be necessary if the earth was to fulfill its promise as a spiritual rehabilitation facility.

> *The one then surviving in the earth, through mineral, through plant kingdom, through the vegetable kingdom, through the animal kingdom, each as the geological survey shows, held its sway in the earth, pass from one into the other; yet man given that to be lord over all, and the ONLY survivor of that*

creation. [Edgar Cayce reading 900-340]

There are several readings that assign a specific time frame for soul activity or incarnations on the earth before about 12,000 years ago (Cayce, 1968). They are reviewed herein primarily to compare with current scientific ideas about the evolution and history of mankind. The accuracy of these readings is difficult to ascertain. Some of these readings are confusing and difficult to interpret or reconcile with current ideas in anthropology and geology. In response to a question about how the second ruler in Egypt gave the first laws that concerned the relationship between God and man to the people, paragraphs two through four of reading 5748-1 begins by describing a period of pre-Egyptian history "so that the conditions of the earth's surface and the position of man in the earth's plane be understood." These readings indicate that the boundaries of the African continent once were approximately coincident with the Sahara and Nile regions, and that the positions of Asia, Europe, South America, and western North America were in similar positions as we see them today. However, the Nile River discharged into the Atlantic Ocean and some other river systems of the world did not flow in the directions we see today. Paragraph three of reading 5748-2 picks up the narrative of earth changes again, stating that, "The period in the world's existence from the present time being ten and one-half million (10,500,000) years, and [with] the changes that have come in the earth's plane many have risen in the lands." It states that many lands have appeared and disappeared during these periods. The four paragraphs in these two readings that describe earth changes (5748-1 pp2, pp3, pp4, and 5748-2 pp3) do not seem to be related to the remaining text in those readings.

Paragraph four of reading 5748-1 mentions that at some time in the future, "man appeared in five places then at once—the five senses, the five reasons, the five spheres, the five developments, the five nations," which we will see in a later paragraph refers to events about 106,000 years ago. Other paragraphs of reading 5748-1 tell how the second ruler brought peace out of the chaos brought on by invaders from the Tibetan and Caucasian regions, and 5748-2 relates that "in the earth's plane at that period, the numbers then of human souls in the earth plane being a hundred and thirty and three million (133,000,000) souls," presumably at the time of the second ruler in Egypt about 10,500 BC. Reading 5748-2 also indicates that the law was formulated into the Book of the Dead, possibly as the first laws to codify the study of the individual body and mind, and the relation between the soul of man and the various planes of existence it can encounter after the death of the body. Modern scholars believe the Egyptian Book of the Dead was compiled by scribes over a long period of time starting from about 1,500 BC and is partly based on older works from as early as 3,000 BC. The book is interpreted as a compilation of magic spells designed to assist the dead in their journey through the afterlife.

These descriptions of changing landforms and a time frame of 10.5 million years in the past are completely out of context with the reference to the appearance of five races and the social, religious, and political activities in Egypt. These landform conditions and changes do not fit with the probable geologic condition in Egypt about 10,500 BC. The presence of the Nile River in lower Egypt at that time was probably the root cause of the Egyptian civilization and the construction of the pyramids in that locality. Earth changes were occurring in Atlantis at about that time, but that continent was not mentioned in regards to these changes. Apparently, the source of the readings intentionally digressed and inserted a brief description of geologic conditions 10.5 Mya, but there is no reason to assume that this geologic digression infers that there was soul incarnation or an active civilization in that distant past. Recent scientific studies of the Nile River suggest that the river has flowed northward for at least 30 million years because the higher elevation of its source area is controlled by a persistent upwelling convection cell of magma beneath the crust of the East African Rift zone (Faccenna and others, 2019). Previous investigations concluded that the uplift did not form until about 6 Mya and before that time the Nile River flowed westward from its source area through Central Africa toward the Atlantic Ocean. The readings would support the conclusion of the older studies.

Reading 2665-2 is more direct and describes a previous life about 10 Mya "when the first peoples were separated into groups as families." The subject of the reading was a female who in a previous life was devoted to making a home and trying to convince the community group to develop and hold to family relationships in what is now northwestern New Mexico (the location of the Mesa Verde cliff dwellings). She left physical evidence of her life in this region in (rock?) drawings that are still visible today. This is a clear reference to soul activity, apparently in incarnated form, on the earth near the time of the Hominini tribe anthropological split (Figure 2). It appears to be the only clear reference in the readings to incarnated soul activity before about 200,000 years ago. It seems to refer to a full soul incarnation and not a looser relationship whereby souls are vicariously experiencing materiality by partially inserting themselves into an emerging pre-Australopithecine species or other pre-human life form. The suggestion of family activity makes the reading more relatable to human behavior as we know it today rather than chimpanzee-like behavior. However, it is difficult to imagine close family relationships, the making of a home, or monogamous activity in pre-Australopithecine or chimpanzee social groups 10 Mya.

Several other readings mention southwestern America as centers of activity in the early history of man but not in the context of community activity 10 Mya. Reading 816-3 tells about a previous life of an individual who was in southwestern America in those portions that were "the place of refuge from Mu and the upper or first activities of the Atlantean land—and in that now known as Arizona and Utah did the entity then become the first of those that established the cave dwellers, and the use of the metals as a medium of exchange and adornment." The reference to cave dwellers may be the ancestors of the ancient Anasazi cliff dwellers of that area. The "first activities of the Atlantean land" would seem to refer to a time before the first period of major destruction in Atlantis that also happened much more recently than 10

Mya. Reading 851-2 tells of a person who was in southwestern America "during those periods when there were the changes that had brought about the sinking of Mu or Lemuria, or those peoples in the periods who had changed to what is now a portion of the Rocky Mountain area; Arizona, New Mexico, portions of Nevada and Utah," and indicates that she was active in teaching the Law of One. This activity could only have occurred after the entry of Amilius into the earth about 106,000 years ago (see below). Other readings suggest that the sinking of the continent of Mu occurred somewhat before and contemporaneously with the breakup of Atlantis, not as early as 10 Mya. The thrust of these readings is that the first soul activity in southwestern America as presented in the readings refers to events that occurred much more recently than 10 Mya and therefore throws serious doubt on the statement in reading 2665-2 that describes activity at about 10 Mya.

Reading 900-426 indicates that [900] believed that [195] was told in a reading that he had lived on the earth 10.5 Mya and was seeking to understand why the next appearance on the earth (12,000 years ago) was given so far in the future. The source of the readings stated that "The appearances in the earth plane are given in full," indicating that there were no other experiences to mention. The reading also stated that the soul had been experiencing other methods of development. A search of the seventy-one readings given for [195] did not uncover any evidence that [195] was told he lived about 10.5 Mya. Reading 195-8 mentions that before an experience in the first Egyptian dynasty [195] was on the earth "in the first days, when the soul's development began in earth's plane" as a worker in iron, and reading 195-14 describes activity "during the first of the appearances of man in the earth's plane" and indicates that this activity was associated with the entry of the five races. Therefore, the reference by [900] to soul incarnation by [195] some 10.5 Mya in reading 900-426 does not seem to be correct. The comments about the first days of soul development and first appearance of man in association with the five races could only mean the period of soul incarnation associated with Amilius, which occurred much more recently than 10.5

Mya but well before 12,000 years ago. Before the entry of Amilius and the many souls that incarnated with him, conditions on the earth were not conducive to proper soul development and mental enlightenment. Almost all previous soul activity was uncoordinated individual selfish activity through thought forms and injection into existing life forms.

In summary, it appears that the readings that suggest or imply that souls incarnated in human form on the earth about 10 Mya to 10.5 Mya are suspect. Two of the readings (5748-1 and 5748-2) seem to generally refer to human activity about 10,500 BC and only incidentally to world-wide geologic changes on the earth that would likely have occurred over a long period of geologic time some hundreds of thousands or millions of years before 10,500 BC. The individual that requested one of the readings (900-426) assumed that a previous reading for [195] had indicated that [195] was active as an incarnated soul 10.5 Mya. This assumption was incorrect. Only one reading (2665-2) clearly expresses possible soul activity and communal and societal behavior 10 Mya, but it seems highly unlikely within the context of an earth occupied by the higher apes and monkeys that preceded the Hominini tribe evolutionary split into separate pre-Australopithecine and chimpanzee lines some 7 Mya when the first humanlike animal began to walk upright. It also would be at least 9 million years before speech had evolved in early man and likely millions of years before the migrations from Mu and Atlantis into southwestern America.

In response to a request for information about the earth's surface at the high point of Atlantean civilization, the source of reading 364-4 asks which perspective should be used to answer the questions; from that of the impact of the entry of Amilius and the worldwide greater spiritual understanding he was attempting to bring into the earth or from the material progress brought through commercial activity in Atlantis and among nations? In either case, the comment was made that various activities that might define the time frame of interest had occurred over a period of 200,000 years. The reading then describes several geologic changes across the world that occurred in that time frame. This reading sets a minimum time span of the soul activity in Atlantis of 200,000 years, which would indicate activity beginning at or before about 212,000 years ago and ending about 12,700 years ago, when the last islands of Atlantis sank into the ocean. The reading actually states that the changes occurred in the final stages of Atlantean history about 10,700, but does not say if that time is BC or before present; it is assumed herein to refer to 10,700 BC.

Currently, anthropological evidence indicates that Early Modern Homo sapiens, the first man with characteristics nearly identical to man today, arose in Europe, Asia, and Africa around 200,000 years ago and coexisted with, but slowly replaced, the Neanderthals over a period of about 170,000 years. By about 30,000 years ago, the Neanderthals were no longer a distinct group. The readings describe soul activity during the time when the Late Archaic Homo sapiens (Neanderthal) and Early Modern Homo sapiens (our immediate ancestors) were the most advanced native human species. Besides the suspect and most likely erroneous references to soul activity about 10 Mya previously described, the very general description of soul activity in reading 364-4 appears to be the only mention of soul activity in human form before the arrival of Amilius.

Man was made in the beginning, as the ruler over those elements as was prepared in the earth plane for his needs. When the plane became that such as man was capable of being sustained by the forces, and conditions, as were upon the face of the earth plane, man appeared not from that already created, but as the Lord over all that was created ... . [Edgar Cayce reading 3744-5]

A reading (364-3) that requested information about the character of the people who occupied Atlantis and their spiritual and physical development brought a response that told of the entering of the soul Amilius into Atlantis "some hundred, some ninety-eight thousand years before the entry of Ram into India." Ram (Rama) is believed to have been born in India about 5,100 BC, so the entry of Amilius would have been about 107,000 to 105,000 years ago. The average value of 106,000 years ago is used herein. Amilius, who entered along with many other souls concentrated in five different widely spaced localities throughout the world, was attempting to bring spiritual enlightenment to the many souls that had lost consciousness of their true origin as souls created by God and who had now become trapped in the material world. Amilius is the soul that entered into human form as the man called Adam, whose story as a person and as an archetype of all the souls who entered with him, is told in the Bible. He was the leader of the red race that projected into the continent of Atlantis. The souls that entered with Amilius into other parts of the world formed the black race in the Nubia region of Africa, the white race in the Caucasian Mountains of western Asia, the yellow race in the Gobi region of central Asia, and the brown race of the western coast of South America (364-13).

All of the souls that entered as part of this mission entered as thought forms and physical body mixtures that later hardened into the physical form that we see today. These bodies were especially designed by God for this coordinated soul activity. The form and structure of these bodies were patterned after existing human life forms, the Homo sapiens species that were already present and active on the earth. They were created outside of the ongoing evolutionary process independent from the similar existing human forms that were part of the natural world and were designed to be a distinct and spiritually pure race. This seems to be the second major intentional intrusion of God into the natural evolution of the universe, the first being the creation of biological life from inanimate matter. The descriptions in the readings of this Adamic race as Sons of the Law of One and Sons of God is based on their origin as a separate category of beings in human form with spiritual purpose. Their spiritual purity came from the fact that they previously held true to God as souls in the spiritual realm and in their willingness and desire to remain true to God's plan when incarnated on the earth. They were not part of the group of rebellious souls that had disrupted the order of the spiritual realm. The time of entry of these souls into the earth and the new body form they used would seem to indicate that there was no need for God to look to the higher apes for inspiration and pattern. There were already higher-order human life forms living in societal groups that could communicate with one another.

When the Creative Forces, God, made then the first man - or God-man - he was the beginning of the Sons of God. Then those souls who entered through a channel made by God - not by thought, not by desire, not by lust, not by things that separated continually - were the Sons of God, the Daughters of God. [Edgar Cayce reading 262-119]

The opposing forces were called the Sons of Belial, those souls that did not aspire to lead lives of spiritual purity, that had little desire to keep true to the laws of God, and that were trapped on the earth and its surroundings because they were enamored with the sensual attractions of materialism and had lost awareness of their creator. Many of these souls initially had entered the earth as thought forms, bodies that perhaps mimicked and interfered with the existing and naturally evolved early Homo sapiens, but which were not fully solidified into material forms. They are described as originally having the ability to separate into male and female forms in a manner similar to the way an amoeba can manipulate the shape of its body (364-3). As these more insubstantial bodies were used to gratify selfish and sexual desires, they slowly became hardened into solid matter and the mental desires of the souls kept drawing them back into the earth. Souls who were manipulating and merging with plant and animal forms may have just as easily been injecting themselves into and controlling the existing Neanderthal and Early Modern Homo sapiens populations to experience sexual gratification through their bodies. Homo sapiens bodies under the influence of the lost souls and their soul-directed descendants could well have been the Sons of Belial with which the followers and descendants of Amilius had been forbidden to physically intermingle.

Before that we find the entity was in the Atlantean land, during those periods when there were the divisions between

> *those of the Law of One and the sons of Belial, and the*
> *offspring of what was the pure race and those that had*
> *projected themselves into creatures that became as the sons of*
> *men (as the terminology would be) rather than the creatures*
> *of God. [Edgar Cayce reading 1416-1]*

The newly created Adamic body form was better suited to the purpose of allowing souls to express themselves in material form and was a better vehicle for soul incarnation and spiritual growth. It is possible that the new bodies were specifically designed to replace existing Homo sapiens bodies that were naturally evolving. These souls and the new bodies they occupied were not supposed to mingle physically with the existing human life forms on the earth. This directive did not seem have anything to do with race consciousness or segregation but was supposed to keep the purity of purpose undefiled so that the incarnating souls did not become contaminated with or descend into the carnal selfishness that had overwhelmed the trapped souls. The plan failed as the newly incarnated souls and their descendants became enamored with the offspring of the Sons of Belial and intermingled sexually with them for carnal pleasure. The mixed offspring resulting from the banned mating between these two groups were called the sons and daughters of men as opposed to the Sons and Daughters of God, which were the pure offspring of the Adamic bodies resulting from interbreeding among themselves (1416-1 and 5245-1).

> *… there may be added that viewed by the re-iterating of the*
> *story of Eden in its beauty, for the comfort of MAN'S indwell-*
> *ing; man's rebellion and consorting with others for aggrandiz-*
> *ing of SELFISH motives FIRST brought Satan, or the serpent,*
> *into the Eden. [Edgar Cayce reading 3976-9]*

The Sons of Belial opposition group, directed by selfish material-oriented souls, must have been a close species relative of the created Sons of God, or biological mating between them would not have produced offspring. The Neanderthals and Early Modern Homo sapiens were the primary human forms with which the Sons of God might have interbred about 100,000 years ago. One can imagine that God used the Neanderthal body form and/or the Early Modern Homo sapiens body form as the pattern to create Adamic bodies and that the Adamic bodies that comprised the five mass projections and their descendants were only a slight genetic modification of the natural evolved forms. The physical forms of Neanderthals probably can be distinguished in the fossil record from the first Adamic bodies, but the physical forms of Early Modern Homo sapiens would be more difficult to distinguish from the first Adamic bodies in the fossil record. But could the abrupt appearance of even a large number of Adamic bodies on the earth within a small interval of geologic time really be detectable in the sparse fossil record? Because of genetic mixing, descendants of the first Adamic bodies would have soon become virtually indistinguishable from the Modern Homo sapiens, who are believed to have appeared about 200,000 years ago. Will scientists ever find sufficient fossil evidence to verify or refute the readings account of the injection of a new body form throughout the world about 106,000 years ago, or are the genetic differences between the new and existing body types too small to detect in the fossil record?

The ability to speak is correlated in the readings to a higher vibrational state and is associated with the pinnacle of evolutionary development in the animal kingdom. The wording of reading 294-11 indicates that speech is a product of evolution and was therefore probably a characteristic of the Neanderthal and Early Modern Homo sapiens, and was not an aspect of man that began with the entry of the Adamic bodies. If souls inserted themselves into evolved human bodies and manipulated muscles for hunting and other life activities, in theory they should have been able to manipulate the existing muscles of the throat and the associated hyoid bone to form speech. That would also make sense in the context of interbreeding in that it would seem less likely for there to be a widespread tendency for souls in Adamic bodies to mate with an evolved mute human species. Mating would be more likely to occur with a human species with which they could communicate.

… speech is the highest vibration that is reached in the animal kingdom, and in that respect man in his evolution is above that of the other creatures in the creation. [Edgar Cayce reading 294-11]

One might assume that the Cayce concept of five races cannot be compatible with science unless genetic studies verify that there were five separate genetic points of origin for the human species, but this would be an incorrect assumption. Reading 5748-2 gives a human population of 133 million at about 12,500 years ago. We don't know the population at about 106,000 years ago when the Adamic bodies were formed but could take a wild guess at ten million. We don't know how many Adamic bodies were thought formed or how long it took for them to solidify into physical bodies capable of interbreeding with existing humans. Assuming that each of the five racially distinct Adamic body groups totaled 10,000 individuals at each of the five locations, Adamic bodies would only constitute about 0.5 percent of the total population of the earth. This would be a genetic drop in the bucket. There is no reason to believe that the Adamic forms numerically overwhelmed the existing human populations. The genetic impact presumably would not have been sudden, given the thought form nature of the original Adamic bodies and the directive to remain pure and unmixed from the existing population, but would have occurred gradually as thought forms solidified and later generations began to violate the purity rule. It should be stressed that the purity rule was not a racial segregation rule in the sense we might think of it today but was a spiritual rule to keep sexual gratification and loss of God consciousness from diluting and interfering with the purpose of the arriving souls and the ever-present danger that the original God awareness of the Adamic souls could be compromised.

There is also no reason to believe that the genetic code of the five injected races was significantly different from the general population into which they were added. The physical characteristics of each of the five races could have closely mirrored the general population into which they were injected. Skin color differences could have already existed. Indeed, the readings suggest that skin color differences among these groups developed in response to environmental factors before and after the soul influx through Adamic bodies. Scientists believe that white skin pigmentation (or lack of pigmentation) resulted from genetic adjustment to the relative lack of sunshine in more northern climates, and black pigmentation was a defensive adjustment to the intense solar radiation of the tropical regions. The readings suggest a similar concept. Reading 364-3 indicates that the transition of Adamic bodies from thought form to solid physical objects was accompanied by color changes that were induced by the surrounding environment of the bodies "much in the manner of the chameleon." This doesn't mean that they could alter their color at will, although as a pure thought form that may have been possible. This is an expression of the evolutionary concept that biological organisms alter their body structure to adapt to environmental conditions, with the additional complication that the Adamic thought forms were probably being mentally modified as they were undergoing the transition from a pure thought form into a body form not readily manipulated by the mind. Science does not have to genetically confirm that man arose out of five distinct genetic groups to be compatible with the readings. The genetic code in modern humans is almost all derived from the genetic code of non-Adamic humans. About 98 percent of that genetic code is identical to the original genes of the Hominini tribe at the time of the pre-Australopithecine and chimpanzee genetic split about 7 Mya and is shared by humans and chimpanzees today.

There is no definitive scientific evidence resulting from ancient DNA analysis that mankind descended from a single pair of original humans who first appeared on the earth about 6,000 years ago. There is also no scientific evidence that mankind descended from an original pair of humans or a large group of similar humans who first appeared on the earth 106,000 years ago. Ancient DNA studies do confirm that modern human DNA contains a small but significant mixture of Archaic Homo sapiens DNA which verifies that the ancestors of modern humans interbred with preexisting and co-existing species like the Neanderthals. As scientists develop more advanced techniques for analyzing ancient DNA, they are better able to reconstruct the genetic code in DNA fragments extracted from ancient fossils. In 2020, scientists successfully obtained DNA from an 800,000-year-old fossil tooth belonging to a human species thought to be the last common ancestor of the Homo sapiens and Neanderthals (Welker and others, 2020). Human life forms were active on the earth long before creationists claim the earth was created and long before the readings state that Adamic bodies came into existence. Ancient DNA studies do not conflict with the assertion of the readings that modern humanity descended from Adamic body forms that were introduced into populations of human life forms that were already active on the earth.

Reading 877-26 discusses an incarnation on the island of Poseidia; one of the five main islands that comprised Atlantis after a major volcanic catastrophe fractured and divided the continent. In response to a question about whether [877] had known Edgar Cayce at this time, he was told that his shared experience with Cayce went back much farther than a few thousand years and in fact "this sojourn was nearer to fifty or five hundred thousand years before we even have the beginning of the Law as the Law of One manifested." The manifestation of the Law of One signifies the entry of Amilius into Atlantis about 106,000 years ago. If we assume the source of the reading meant 50,000 years, the time frame of about 156,000 years ago would fit within the period from 212,000 to 12,700 years ago mentioned as the major period of Atlantean activity in reading 364-4. This may or may not be consistent with a reference to a Gobi experience mentioned in reading 877-12, which mentions the Gobi (Mongolia) region in the context of interaction with the people of Atlantis, Mu (Lemuria), India, Egypt, Caucasia, and Pyrenees. If we assume 500,000 years is correct, the time frame of 606,000 years ago seems too far in the past for such organized international activity on the earth. It could refer to shared activity in the spiritual realm, but why associate it with an earthly time frame? Perhaps the time frame of 50,000 years is a better fit, but there seems to be no way to verify it. At best, it is difficult to interpret the meaning of these time frames and the type of shared activity to which they refer.

The readings indicate that at some point in time the emerging human societies were beginning to feel threatened by the presence of the large numbers of huge animals that roamed the earth. The threat was so serious and widespread that several nations felt the need for a concerted and coordinated effort to eradicate them or at least reduce their menace to mankind. A council of nations was called by Atlantean rulers in Poseidia for the purpose of organizing and coordinating the effort in 50,722 BC (262-39; incorrectly given as 50,772 BC in Cayce, 1980). The chosen method to do this on a large scale was to alter or change the environment needed for the sustenance and survival of the great beasts in selected regions of the earth where they dominated nature. Presumably the threat was greatest in the more northern latitudes near the southern boundaries of glaciers where massive creatures such as mastodon, giant elephants, Irish elk, and many others roamed freely. The mentioned time of this organized effort would have occurred toward the end of Neanderthal activity, well within the period of activity of Early Modern Homo sapiens, and about 55,000 years after the influx of souls into Adamic bodies. There is ample fossil evidence that the Homo sapiens species hunted these large game animals but no scientific evidence to indicate that they felt threatened by them some time during their societal development.

Readings 364-4 and 364-6 also describe a concerted effort to eliminate large animals, apparently through the use of explosives at the places where they tended to congregate. This worldwide hunt is described in reference to the activities of Atlanteans. These readings suggest that the threat posed by the animals coincided with the first major destruction of Atlantis, which resulted in the large continent being broken into several large land masses separated by water. It is possible that the continental destruction caused the animals and humans to become concentrated in the surviving land masses as they sought safety and forced them into closer proximity and caused more confrontations nearer to the human habitations. This activity apparently refers to the same time period of 50,722 BC given in reading 262-39. The continental destruction was by geologic activity, but the readings indicate that the failure of humans to hold true to the purity of purpose embraced by the Sons of God was the underlying cause of the destruction. The use of explosives in this effort eventually lead to misuse of the same forces in ways that brought further destruction on Atlantean society and the continent of Atlantis, eventually leading to mass migrations into various regions including migrations into lands that would become the centers of Egyptian, Incan, and native North American activity.

Reading 5748-3 mentions that Egypt was relatively free of the threat to life and property from wild animals but that it was still a problem in Asia, Europe, and the northern and western portions of the earth, which may indicate Atlantis and the elevated areas of western North America, which were dry land then. The reading mentions a successful effort to control and eradicate the dangerous large animals that overran the lands and threatened the emerging societies of man. The reading indicates that this effort to gather five nations together to control wild animals was organized by the second ruler of Egypt. This would make the time frame about 12,500 years ago, so would refer to activity about the time of the final break up of Atlantis and not to the time of the first break up.

> *Then with the breaking up [of Atlantis], producing more of the nature of large islands, with the intervening canals or ravines, gulfs, bays or streams, as came from the various ELEMEN-TAL forces that were set in motion by this CHARGING - as it were - OF the forces that were collected as the basis for those elements that would produce destructive forces, as might be placed in various quarters or gathering places of those beasts, or the periods when the larger animals roved the earth - WITH that period of man's indwelling. [Edgar Cayce reading 364-6]*

Two successive major glacial periods have occurred during the past 240,000 years in North America and Europe, and they would have strongly affected life in Atlantis. The earlier period began about 240,000 years ago at the end of a previous short interglacial warm period and continued through a long period of stepwise declines in average temperature until about 140,000 years ago. This cold period ended with an abrupt increase in temperatures that initiated a 15,000-year-long interglacial warming period. About 125,000 years ago the most recent glacial period began and followed the same pattern of stepwise declines in average temperature until about 19,000 years ago when another relatively abrupt increase in average temperature signaled the onset of the current interglacial period (National Centers for Environmental Information). The earth is likely at or near the end of this latest interglacial cycle. The coldest temperatures and the years of the most southern advance of the continental glaciers would have occurred in the latter half of the two glacial periods from about 194,000 years ago to about 140,000 years ago and from about 65,000 years ago to about 19,000 years ago. During these extreme conditions, about one-half of North America, Europe, and Atlantis would have been covered by huge ice sheets. The humans and animals (large Pleistocene mammals such as mammoths, mastodons, and saber-toothed cats) that inhabited these continents would have been slowly driven into closer proximity in the southern reaches of the continents.

These large mammal species that shared the earth with humans during the last ice age had survived multiple glacial and interglacial periods of climate change for millions of years, but most of these large species went extinct about 12,000 years ago. The current scientific explanation is that they were hunted to extinction and did not die off as a result of climate change. Increased hunting by early man and loss of habitat as man encroached on their ranges is believed to have led to a substantial decrease in the numbers and distribution of these animals, which finally drove them to extinction. The rapid pace of the extinction is unusual for such a large and diverse group of animals and is one factor suggesting that the extinction was caused by human intervention and not natural causes.

A study by Faith and Surovell (2009) was undertaken to determine whether temporally staggered events or an abrupt and catastrophic event caused mega faunal extinction. They concluded that although the fossil record presents challenges for the precise calibration of the extinction chronology, there is quantitative evidence providing a clear support for an extinction chronology of North American Pleistocene mammals as a synchronous event that took place 12,000 to 10,000 radiocarbon years before present. Their results favor an extinction mechanism that wiped out 35 genera across the continent in a geologic instant. It is diffi-cult to reconcile the fossil evidence of abrupt mega faunal extinction 12,000 years ago with the readings time frame of 50,722 BC (364-4, 364-6, 262-39) for a concerted effort by multiple nations to exterminate mega fauna unless one imagines that the effort recorded in the readings did not exterminate the beasts but only ridded the nations of beasts that were an immediate threat to communities, leaving the remaining population of large animals to be reduced over a longer period of time by increased hunting and habitat degradation. If the 50,722 BC time frame is correct, it must refer to a previous eradication attempt that scientists have not identified or was too small to be identified. However, the eradication effort mentioned in reading 5748-3 is consistent with the complete destruction of the animals that scientific studies suggest occurred within a short time span about 12,000 years ago.

Reading 3517-1 described a serious and painful physical ailment caused by irritation or pressure within the cerebrospinal system of the subject of the reading. A follow-up reading stated that the origin of the physical problem was rooted in previous activity from about 35,000 years ago and suggested that relief from the condition could be found by continuing with recommended treatments and having trust and faith that a divine influence was actively working in her body. The place of the previous incarnation was not given, but the time frame is consistent with the period between the first and second major land disturbances in Atlantis when the Neanderthal was at the end of its existence as a separate species and when science indicates that the mixing of genetic material between the Neanderthal and modern man would have been nearly complete. The reading suggests that Adamic bodies were well on their way to dominance by this time.

Information about the second in a series of three widely spaced major cataclysmic events in the history of Atlantis that eventually resulted in its total destruction by submersion in the Atlantic Ocean is recorded in several readings. Readings 2625-1 and 470-22 provide information about two individuals who lived on the earth about 28,000 BC (30,000 years ago) during the second period of major activity relating to the migrations from Atlantis and attempt to preserve lives and records from the pending geologic catastrophes. These persons were related to Japheth, a son of Noah, and were present during and helped with the preparation of the ark (Genesis 6–10). This geologic event was the inspiration for the story of the flood recorded in the Bible. These readings define a time frame for the biblical Great Flood and describe the flood within the context of continental upheaval, subsidence, and large-scale migration from affected areas during the second breakup of Atlantis. Reading 364-6 indicates that the second period of destruction occurred about 24,000 years ago. To reconcile this time frame with the earlier time given in readings 2625-1 and 470-22, it must be assumed that the Atlantean continent was periodically coming under threat and that migrations from the continent were scattered in time and did not occur as one massive migration within a short time frame. The time frame of the second destruction of Atlantis would have been about 80,000 years after the influx of Adamic bodies when the Neanderthal was nearly extinct as a separate species and Early Modern Homo sapiens was nearing the end of its reign. Incidentally, reading 470-22 also gives the time of the exodus of the Israelites from Egypt as 7,500 years ago.

The Epic of Gilgamesh, which describes a Chaldean flood, dates to about 2,700 BC, or 4,700 years ago. The actual time frame of the flood is unknown, and the described flood can be interpreted as either a local river flood or as a worldwide deluge. A bracketed note in paragraph three of reading 311-3 assigns the time of this flood to 10,000 BC, which would likely refer to the final breakup of Atlantis. Using the scholarly interpretation of the epic, the Chaldean flood describes a much later event than the Noah flood, which is associated in the readings with the second breakup of Atlantis (364-6). The words "twenty thousand years before the Chaldean message, or flood" in reading 311-3 can then be interpreted to mean that the subject of the reading lived a previous life about 25,000 years ago. The reference to a life in a country "now called the Indian land" and the place names given in the reading suggest the Indian subcontinent. This incarnation apparently occurred just before or during the second major land disturbances in Atlantis, but half a world away from that catastrophe, and shortly after the last of the Neanderthals existed as a separate species. An earlier incarnation is also given for [311] in the American land, but no time frame is given.

In response to a question about the cause of the destruction of portions of Atlantis now under the Sargasso Sea, reading 364-11 states that the destruction followed from the misuse of forces arising from discoveries relating to gases and solar radiation. The Sargasso Sea is a region of slow-moving somewhat stagnant water in the Atlantic Ocean off the coast of Florida and the Carolinas bounded by the clockwise circulation of several Atlantic Ocean currents including the North Equatorial Current, the Antilles Current, the North Atlantic Current (Gulf Stream), and the Canary Current. It extends over one-third to one-half the distance between North America and northern Africa. The region would have constituted the southern or southeastern portion of Atlantis. The date of this period of turmoil was given in relative terms as 7,500 years before the final destruction of Atlantis. A previous publication (Cayce, 1980) assigns an approximate date of 8,150 BC to this final destruction, but this date seems to be much too recent. The final destruction of Atlantis is connected in several readings to the periods of exodus from Atlantis to Egypt ranging from about 12,500 BC to about

10,500 BC (Cayce, 1968) rather than a particular year being defined in any single reading on the subject. Assuming the beginning of the final destruction of Atlantis began about 12,500 BC, the period of destruction to which reading 364-11 refers would have occurred about 20,000 BC, or 22,000 years ago, and is consistent with the commonly accepted second destruction of Atlantis. No geologic evidence has been discovered that supports the destruction of a continental land mass now buried beneath the Sargasso Sea.

One reading (288-6) describes the activity of two individuals in aiding the spiritual and moral development of the Atlanteans. They were associated with the religious and legal system of Atlantis about 12,800 BC (14,800 years ago) and their work was complicated by strong sexual desires. One of the individuals was told again in a later reading (288-39) that excessive sexual attraction and desire in the life in Atlantis was the cause of a self-destructive life and needed to be replaced with a strong desire to seek Oneness with the Creator. The advice given to the individual applies to all incarnating souls. In response to a question about man-animal mixtures in 12,500 BC, reading 2067-6 verified that souls were still involved with animal forms during that period and that it usually took several generations for the appendages associated with those mixed forms to be cleared from the offspring. The presence of remnant plant and animal physiological features were still seen in some people as late as about 10,500 BC when facilities were constructed in ancient Egypt to complete the removal of nonhuman appendages.

There are many readings that mention the prevalence of thought forms or "things," souls embedded in partly physical, partly nonphysical, humanlike forms that were looked down upon by others and often were treated as sub-human and usually forced into slavery to serve the needs of others. These thought forms are described as being synonymous with evil, meaning that their condition was a direct consequence of rebellion in mind against God and misuse of spirit for self-gratification (585-12, 1472-10 and 2072-8). The individual who was the subject of reading 2625-1 and who lived about 30,000 years ago also incarnated in Egypt about 12,500 BC (14,500 years ago) and was active in bringing together and merging the ideas and purposes from Atlantean and Egyptian religious traditions. The time frame of these several readings associated with the final breakup and sinking of the remaining islands of Atlantis marks the end of Early Modern Homo sapiens and the transition of humanity into the Late Modern Homo sapiens species that forms all human populations today. From the perspective of the readings, Late Modern Homo sapiens represents the final expression of the Adamic body type mixed with identifiable genetic remnants of Neanderthal and Early Modern Homo sapiens. It took about 95,000 years for all souls to cease activity on the earth as thought forms and cease entanglement in the plant and animal kingdoms. All future soul incarnations would occur through the Adamic body type.

In a comprehensive overview of dramatic new scientific developments in the genetics of the human population, David Reich, professor in the Department of Genetics, Harvard Medical School, discussed unexpected genetic mixtures of North American Indians in the current European population (Reich, 2018). In trying to understand how that mixture occurred, the lecturer makes the statement in passing that, of course, the Native Americans did not have the ability to cross the Atlantic Ocean in the numbers required to account for the mixture. The scientific community has not made a serious effort to verify the existence of the ancient continent of Atlantis, but the existence of that continent and the mass migrations away from the sinking continent during periods of geologic upheaval described in the readings would easily explain the observed distribution of Native American DNA in the cultures of North America and Europe. To explain the observed mixing, the researchers had to resort to a statistically derived ancient ghost population that traveled across Asia and the Bering Strait into North America at least 15,000 years ago and that has since died out. This hypothesis is weakly supported by one 24,000-year-old fossil sample from a boy found in Lake Baikal, East Central Siberia. The DNA from this sample has a strong similarity to Native Americans and Europeans and little similarity to modern Siberians.

When describing the DNA of ancient fossils discovered in the Americas, the lecturer describes three older genetic sequences from a 10,600-year-old sample from Chile; a 9,600-year-old sample from Brazil; and a 7,400-year-old sample from Belize that share common traits but are not closely related to modern populations in those areas. However, they are closely related to a Clovis culture DNA sample from southwestern North America. The lecturer assumes that this signals an ancient migration of Clovis population members southward but not to the extent that the Clovis culture was transferred into the southern regions. Again, these common DNA characteristics in approximately 10,000-year-old ancient fossils could also be explained by the mass migrations of Atlanteans into North America, Central America, and South America during the second and the third destructions of Atlantis. In a presentation of current knowledge about the observed genetic mixing of Neanderthal and associated populations with modern humans, the period between 54,000 years ago and 49,000 years ago is identified through genetic trait dilution as a time of major interbreeding among human species. The lecture is filled with references to recent publications in human genetics that are revolutionizing the current understanding of the migration and mixing of pre-historical human populations and the reader is encouraged to become acquainted with these studies.

Physical Consciousness And Causality

There is no time: it is one time; there is no space; it is one space; there is no force; other than all force in its various phases and applications of force are the emanations of men's endeavors in a material world to exemplify an ideal of its concept of the creative energy, or God, of which the individual is such a part that the thoughts even of the individual may become crimes or miracles, for thoughts are deeds and applied in the sense that these are in accord with those principles as given. That that one metes must be met again. That one applies will be applied again and again until that one-ness, time, space, force, or the own individual is one with the whole, not the whole with such a portion of the whole as to be equal with the whole. [Edgar Cayce reading 4341-1]

Typically, a soul is attracted to a newborn body when the present and potential future of the physical, mental, and spiritual environment of the child will provide the opportunities and instructions needed for that soul to grow spiritually. In some instances a soul may be directed to a particular infant's body and environment instead of having a conscious choice because the soul's degraded condition requires a particular collection of material and spiritual circumstances. One might think that God is ignoring that soul's free will, but that is not correct. This situation might be analogous to a person who jumps off a cliff with the intention of falling upward. The person may want and hope to fall upward but the physical laws that control the motion of matter will demand that the body of the person fall downward. Spiritual laws are in effect whether or not the soul understands them or acknowledges them. Every soul creates negative karmic conditions when it fails to make spiritually correct choices while occupying a physical body. Spiritual laws require the soul to again and again meet a situation similar to that one in which it previously erred. The soul can't decide it doesn't want to face its shortcomings and spiritual failures, just as a person can't decide he doesn't want to fall downward in the earth's gravitational field at the edge of the cliff.

When a soul choses a newborn body of its own volition, its choices are still limited to those bodies born in environments that will provide the proper conditions needed by the soul to meet some aspect of its mental and spiritual self that needs to be strengthened, corrected, or redirected toward God consciousness. It may be the case that a newborn body provides those conditions for more than one soul that is ready to enter the earth, and that one soul may lose that particular opportunity to another soul. A soul seeking a new body also may be confined to or directed toward the region where its previous body died instead of being able to look worldwide for a new body. This may provide continuity of physical experiences and interpersonal relationships, the opportunity to again meet people (other souls) with whom they previously interacted or with whom they had a close physical relationship or had a specific personal problem that needs to be faced again. Dr. Ian Stephenson, a noted investigator of reincarnation claims, studied twenty-four instances in which children living in Burma (Myanmar) claimed past-life memories as Japanese soldiers who died in Burma during World War II and who were subsequently reincarnated in Burma, often with the feeling that they didn't belong there (Stevenson and Keil, 2005). These souls may have been required to face the people and country that they conquered and occupied as soldiers.

Our physical conscious mind experiences the world of atomic matter through the concepts of time and space. Before the soul links with a human infant body, it has no concept of time and space and is largely unaware that its thoughts and activities can cause it to become separated in consciousness from God. The awareness of time as an ordered series of events allows the mind to recognize the physical changes that it causes in the world as it exercises its creative abilities and to assess the mental and emotional repercussions of events it sets in motion that affect other souls. Physical consciousness is rudimentary in an infant. It develops as the infant consciousness absorbs and catalogs sensory information collected by the sensory organs of the body from its surrounding environment and assimilates the data into a usable conceptual image and understanding of the nature of its surroundings. Part of this understanding is the concept of causality. The soul, through the activity of the conscious mind, begins to realize that every action in the physical world has a consequence, an effect on matter, an effect on the consciousness of other individuals, and an effect on itself. A more difficult concept for the conscious mind to understand, but which can eventually be realized, is that the consequences of spiritually inappropriate events that occur in one lifetime are carried within the soul mind through the dimensions of consciousness that it experiences between earthly lives and must eventually be faced by the soul in subsequent earthly incarnations. This subconscious soul memory is a necessary part of the soul's spiritual requirement to meet itself, to come face-to-face with its spiritual shortcomings. Spiritual error, or sin, committed in the flesh must be met and corrected in the flesh.

> *For, only by aiding others may the soul within itself advance for its development towards filling that purpose for which it came into material experience in the earth … : That there might be manifest in the flesh those things in the mental and soul body that have been gained throughout the sojourns in the earth. For, only by manifested acts that make for a closer relationship of the soul to that source from which it sojourns, may there come the consciousness of self - and in self - being at an at-onement with the Creative Forces or God in the earth. [Edgar Cayce reading 423-3]*

There are over 30 trillion cells in the human body that perform numerous different functions, including epithelial cells (covering all interior and exterior surfaces), nerve cells, muscle cells, and blood cells. Each cell in the human body can acquire its own desire independent and counterproductive to its purpose and function within the body. This dissonance or lack of attunement of the cellular activity within the greater and higher purpose and activity of the body causes dis-ease (uneasiness in soul and physical consciousness) and disease, and endangers the ability of the greater organism to properly express itself and fulfill its true purpose in life. The cell must be healed or caused to cease its disharmonious existence before it severely damages the body. No cell of the human body can survive independent from the body because the body supports the cell by providing it with life sustaining nourishment via the blood system. There is a direct analogy between a sick malfunctioning cell within a human body and a sin-sick soul expressing itself in ways that are counter to its purpose and true nature as a portion of God. Souls are a part of God, the smallest possible units of independent mind-directed spirit, and they cannot remain at odds with their purpose, which is integrated with and part of the greater purpose of God. Like human body cells, souls cannot survive without nourishment, and when a soul blocks the God consciousness that under normal circumstances facilitates a nourishing flow of love to the soul from the creator, the soul cannot function according to its proper role and cannot survive. The sickened soul must be healed to the betterment of the whole or purged from the body of God.

> *For each soul is a corpuscle in the body of God. And when differences arise in a body, where corpuscles are at variance to a common purpose for all, sin enters, and death by sin, to whatever may be that group, that organization, that is stressing differences rather than the coordinating channels through which all may come to the knowledge of God. [Edgar Cayce reading 3395-2]*

Each cell in a healthy body cooperates with the other cells and contributes its particular skill set to the proper harmonious functioning of the larger organism. Just as cells of the human body must cooperate to ensure the proper expression of the whole body in the physical world, so must souls in the body of God cooperate in harmony and peace for the proper expression of God in the physical and spiritual realms. The readings state that we are corpuscles in the body of God [3333-1, 2400-1, etc.]. Each of us has a unique role to play, a set of unique talents and abilities that complement the whole (Romans 12:3–8). Souls that are steeped in the sins of selfishness exist in a state of disharmony with respect to the whole. In effect they are like cancerous cells of the human body that set off on their own way without regard to the damage they are causing to the larger organism. Diseased souls, like diseased cells, must be made healthy again by the correct application of mental and spiritual food to once again take part in the glory of the greater organism. Diseased souls that cannot be redeemed must be pruned from or absorbed by the parent organism for the good of the whole.

> *Then, as the cells of the body are aroused in themselves to that awareness that each cell is to perform a functioning to the glory of a glorified consciousness - not of self but of Him, who is life itself - we may overcome these disturbances. [Edgar Cayce reading 3125-2]*

The mind has been the subject of speculation and investigation at least since the fourth century BC when Aristotle and his followers organized the Peripatetic school of philosophy. As early as the seventeenth century, the philosopher Rene Descartes formulated a dual nature of man in which the material body works in cooperation with an immaterial soul. The body was envisioned as being filled with a substance that travels through the nerves and activates muscles by suffusing them with animal spirits. Today scientists speak of electrical impulses instead of animal spirits. Descartes believed that physical activities of the body, both intentional muscle movement and reflexive muscle response are initiated by the immaterial mind and activated through and controlled by the brain. The immaterial component is only found in humans and is the substance that produces rational thought and consciousness. These mental processes are fundamental properties of every human but are not amenable to the laws of physics. They are carried out and controlled by the soul. The interface between the physical body and soul-derived mind is primarily the pineal gland, the small endocrine gland located in the center of the brain near the root of the spinal cord. The pineal gland regulates the twenty-four-hour cycle of sleep and wakefulness through the production of melatonin, a serotonin-derived hormone that manages sleep patterns.

> *The spiritual contact is through the glandular forces of creative energies; not encased only within the [Leydig] lyden gland of reproduction, for this is ever - so long as life exists - in contact with the brain cells through which there is the constant reaction through the pineal. [Edgar Cayce reading 263-13]*

By the first half of the twentieth century, descriptions of the mind were being formed based on the results of psychoanalysis of mental patients. Sigmund Freud published his theory of the mind in 1933 and described the mind of man as having three major structures; the ego, the id, and the super ego (Freud, 2010). The ego is responsible for self-awareness and partakes of a conscious component that monitors sensory receptors throughout the body and is responsible for perception, reasoning, logic, and the feelings of pleasure and pain. It also has an unconscious component that uses repression and denial to defend the conscious component against the id. The id arises out of the deep hidden unconscious component, which is ruled entirely by thoughts of seeking pleasure and avoiding pain. This unconscious component was postulated to be the only mental structure that exists in newborn infants, the ego being developed as the infant grows into adulthood and accumulates and analyzes experiences. The super ego contains the moral code, ambitions, and urges of the mind. This descriptive picture of the structure of the mind based on psychoanalysis was soon to be displaced by discoveries in the new and exciting science of cell biology that had been slowly maturing during the previous two or three decades. Scientists had discovered that specialized cells called neurons were transmitters of information through the body and were gradually learning the mechanics of their structure and the modes of electrochemical activity that allow them to transmit and process information.

Neurons are specialized cells that collect information from the interface between the body and its environment and transmit that information to the brain. A signal detected at the receptive end of a neuron (dendrite) initiates an electrical pulse that migrates along the body of the neuron (axon) to the opposite transmitting end of the neuron (terminal) where it is passed to the next neuron in the chain by a chemical reaction across the gap (synaptic junction) that separates two neurons. The electrical pulse (action potential) travels about 100 feet per second and is always the same shape and size, no matter the intensity of the stimulus. The intensity of the interaction with the environment (pressure of a touch, intensity of noise or light) is encoded in the neuron as an increase or decrease in the frequency at which the pulses move along the axon. More pulses per unit time indicate a more intense interaction with the environment. Proteins, the essential building blocks of the body, cause and guide the electrical pulse. The outer membrane, or sheath, that surrounds each neuron is embedded with miniature fat protein cylinders that can open or close to allow ions to pass, or prevent ions from passing, through the cell wall and enter or leave the neuron body. Electrical impulses travel along the cell membrane as these protein gates open and close in succession along the neuron wall and ions migrate inward and outward across the membrane and cause the electrical potential across the membrane to change. The location of the signal on the body, the portion of the skin that was touched or the ear drum was that vibrated or the part of the retina on the back of the eye ball that was impacted by a photon, is encoded in the brain at the end of the particular pathway along which the signal is transmitted.

Each neural pathway that starts at a sensory receptor at a specific location on the skin, ear, eye, tongue, and nose ends at a particular location in the brain that only processes signals coming from that location. Sensory cells in the skin absorb and radiate electromagnetic radiation in units of energy called photons and interpret the intensity of the energy gain or loss as heating or cooling. Neurons send most of this information for temporary storage in a column of cells in the parietal lobe of the brain. The constant Brownian motion bombardment of the exterior and some interior surfaces of the body by the atmosphere is largely filtered out and ignored by the conscious mind. When a sound wave strikes the eardrum, the intensity and frequency of information is transmitted to the brain and briefly stored in the auditory cortex. The eyes interpret incoming electromagnetic radiation as vertical, horizontal, and angular boundaries of light and dark and store this information along with frequency and intensity information in the visual cortex. The tongue and nose add to the vast quantity of electrical impulses arriving every second at the brain through the neurons. This information is stored in the taste cortex and olfactory cortex, respectively. The signal being transmitted along one neuron can be passed to more than one receiving neurons in a cascading pattern. Each signal that a neuron is carrying can be tracked back to a specific small area on the surface of the body or within an eye. The brain can determine the magnitude and location of the body-environment interaction and make the appropriate response by initiating a series of signals in a similar network of neurons that propagate signals away from the brain toward the muscles. The brain's response to a warm object touching the skin would cause the brain to direct a small number of action potentials per second toward the appropriate muscles to move the body, and a hot object would cause it to direct a much larger number of action potentials per second through the same neurons to move the body more quickly (Kandel, 2006).

Current scientific research on the mind is focused on the concept that the conscious mind and the behavior of a biological organism is a result of neurobiological activity within the brain. The mind is perceived as the product of biological processes and the goal of scientific research into the mind is to understand how the electrical activity of the brain causes the body to develop self-awareness. The biological structure of the brain can be studied using a variety of techniques such as dissection, response to electrical stimuli, magnetic resonance imaging (MRI), electrical measurements of neuron activity (electroencephalography or EEG), and chemical analyses of neurotransmitters in synaptic junctions. These attempts to study the mind of an individual with electronic instruments are necessarily all undertaken by third-party investigators because the mental awareness of the brain being studied is only available to and accessible by the subject of the study. The brain can be poked and prodded, but the mind cannot be sampled or measured by electronic instruments. No scientist can peer into or insert a probe into the mind of a subject to observe or measure its inherent properties. If the readings are correct in their claim that the cells of the human body and all biological life have a form of consciousness, then the scientific hypothesis that the mind arises from the cumulative activity of neurons is incorrect. Individual cells do not contain neurons.

One question now circulating among neuroscientists is whether the accumulation of physical data that describes the electrochemical processes of the brain is sufficient to give insight into the mental state of an individual, or is the data simply the collective knowledge of physical processes that describe neuron-neuron and brain-body connections that have no bearing on the mind. Kandel (2006) asserts that the discoveries of neuroscience show that the functions of the mind can be explained by the laws of physics and chemistry, and action potentials are the key biological mechanism of the body that convey information about thought, emotions, and memory. Those who believe that the mind can be discovered and quantified by the investigation of the biological processes of the neurons that compose the brain argue that we just need more accurate data at a finer detail and in greater depth, perhaps requiring a quantum mechanical approach to brain analysis. They feel that in some way the physical activity of the brain is the mind, but we do not yet have enough knowledge and understanding to put it all together into a coherent picture (The Hidden Mind, April 2002).

> *... the body, the mind, the soul are one. We see the body with those attributes. We can conceive of what the mind is, or see it in action, but never find the mind, in the body. [Edgar Cayce reading 5254-1]*

The transition from psychoanalytical descriptions of the mind to descriptions of the mind built around neuron function and brain activity moved the quest for the understanding of the mind from nonbiological to biological origins. A purely biological approach does not fit well with the description of the mind that can be inferred from the readings. The primitive pleasure-pain oriented infant mind of Freud bears little resemblance to the veiled but fully cognitive mind of a newly incarnated soul in an infant body with its multiple physical and spiritual experiences as described by the readings. The appearance of the mind as arising out of a vast neural network of electrical impulses along axons and chemical transfers across synaptic junctions in the 86 billion neuron brain (Herculano-Houzel, 2009) cannot be reconciled with the mind as a gift of God created in the image of the Mind of God. If the human mind was irrefutably proven to be purely a biological phenomenon, then we would have to abandon any thought of the continued existence of the mind after physical death, discard the concept of a conscious afterlife that is central to the theology of the major religious, and conclude that the descriptions of the mind in the readings are false. Some scientists today may be ready to make that declaration, but the evidence does not yet support such a sweeping conclusion.

But the neuron theory of the mind does have the potential to bridge the gap between science and the readings. No measurement of brain activity has yet absolutely identified and verified the mind as arising from the biological function of neurons, and the readings suggest that is not likely to happen because the mind is not biological in nature. However, there may come a time when science will determine, detect, or infer that some as yet unknown force is acting on or interacting within the brain and spinal column to induce electrical impulses and perhaps will detect that there are similar yet unexplained interactions occurring within the endocrine glands. Until such observations and measurements are made and initiate another paradigm shift in neuroscience, the scientific understanding of the mind and the explanation of the mind presented in the readings will remain far apart. The descriptions of the mind and its interface with the human body in the readings seems to have more in common with the seventeenth-century speculations of Descartes than it does with the twentieth-century theories of Freud and current trends in experimental neuroscience.

> *... the study of electrical energies is the basis for finding in the scientific manner the motivative force of animation in matter.*
> *... For Life is, and its manifestations in matter ARE of an ELECTRONIC energy. [Edgar Cayce reading 440-20]*

Like the description of the mind advanced by Descartes, the readings indicate the pineal gland plays a prominent role at the interface between the mind and body. The readings often depict the mind as the builder, that part of the soul that constructs our future under the direction of our will. The pineal gland is also called the builder because it guides the development of the brain in a fetus. The suggestion is made that the gland is instrumental in directing nutrients supplied through the umbilical cord into an orderly organization of brain matter based on instructions in the genetic code of the parent's DNA. By the time the umbilical cord is severed at the birth of the infant, the pineal gland has become the control center for a developed physical brain and is the primary interface with the mind of the incarnating soul. The gland takes on the functions that coordinate impulse and imagination in the new body. By impulse is meant the basic urges and instincts that drive the body, the inner compulsion to behave in a certain manner or respond to certain stimuli in a certain manner. By imagination is meant the creative powers of the mind, that from which new insights arise, the ability to be resourceful and original in thought. Damage or deterioration of the pineal gland or its neural connections with the upper portion of the spinal cord can result in hallucinations at any age and senility in advanced years. These conditions may be symptoms of the inability of the soul mind to coordinate with the physical body and suggest a potential fruitful direction of study of the mind-body connection. Perhaps future neuroscience research into the role and function of endocrine glands will close the gap between current scientific knowledge and the readings as pertaining to the source of the mind and the connection between the body and mind.

> *... in the fetus as is begun in first of gestation, we find this [pineal gland] may be termed as the Builder. As is seen, the location of same is in the beginning in that of the center or the nucleus about which all of the matter takes its first form, and becomes the brain as is guiding or directing the building of the body as its development in the womb takes place. ... When there has reached that stage when there is the separation of same, the cord [umbilical cord] then being broken, this forms then its own basis in the lower portion of the brain, or cerebellum, and through the medulla oblongata to the central portion of the cerebro-spinal cord ... . [Edgar Cayce reading 294-141]*

The readings clearly fall into the camp that declares the mind does not arise from brain activity. The mind is the active force within God that allows him to direct and shape Spirit to create the spiritual realm, to set the evolution of the physical universe in motion, and to create life within the physical world. The soul mind was patterned on the Mind of God but developed its own individuality through activity and experience. Each soul that was created in the spiritual realm was endowed with a mind so that it could participate in creative acts, be a co-creator with its Creator. The mind of each individual who exists or ever existed on the earth originated in his or her soul and preceded the development of the person's physical body, brain, and personality. The subconscious mind serves as the seat of awareness of the soul, and the super conscious mind serves as the interface between the soul and God. When the soul is not incarnated in a physical body, the conscious mind is not present. When the soul incarnates into the human body, the subconscious mind and super conscious mind are still active, but the mental body also acquires a physical consciousness, the aspect of the mind that interfaces with the human brain and central nervous system of the physical body, and acquires awareness of the physical universe of matter and energy through the body (Figure 1). This allows the mind to sample, analyze, and catalog the electrical impulses transmitted from sensory organs to the brain. The activity of the physical conscious mind largely hinders or suppresses the awareness of the subconscious mind and relegates the perceived activity of the subconscious to the domain of intuition. When the body sleeps the sensory flow decreases markedly, allowing the conscious mind to be largely deactivated while the subconscious mind of the soul maintains bodily life functions and reveals itself during dream states.

Spiritual and moral qualities in human beings are not biological in origin and do not naturally arise out of a biological organism. They do not originate or arise from the physical realm through human brain activity in response to environmental conditions or from a biological survival mechanism but originate in the spiritual realm from God as inherited traits imprinted as a pattern within each mind and available to each soul as a divine driving force of activity. None of the universal laws that control the spiritual or physical realms are dependent upon or changed by our concept of God or our ignorance of God. God is infinite and is not comprehended through the use of the human brain or application of the finite rational reasoning power of the physical consciousness. When we attempt to define God using the conscious mind, we impose limits upon him that prevent us from fully understanding his nature and power. We can improve our conscious understanding of God and help restore our soul's ability to communicate with God by learning and applying some of his attributes or characteristics. As part of their inherent nature, souls acquired the ability to express the attributes of God, which in the physical realm are the same qualities that were demonstrated in the life of Jesus and expressed in the biblical fruits of the spirit. Traits such as love, patience, kindness, longsuffering, and other societal behaviors that show respect and love for our fellow souls are as appropriate and important to souls as to human beings and bring the human mind into conscious union with the soul mind.

Man finds himself a body, a mind, a soul. The body is self-evident. The mind also is at times understood. The soul or the spiritual portion is hoped-for, and one may only discern same from a spiritual consciousness. [Edgar Cayce reading 2879-1]

The influence of the soul mind on the body it occupies and on fellow incarnated souls can be observed. All souls have a common purpose for entering the earth; to awaken the Spirit of God that resides within the Holy of Holies, the seat of the highest spiritual consciousness of man located deep within but also beyond the temple we know as our physical biological bodies. This awakening is a process, not a revelation or salvation moment, and is accomplished by persistent application of the fruits of the spirit, the fundamental moral, mental, and emotional qualities of God. Every relationship and every experience in which we participate while on the earth offers us the opportunity to express these qualities. It applies to close relationships among our family members, acquaintances from work or the neighborhood businesses, strangers we meet on the street, and even those who express a dislike or hatred toward us. It does not mean we can think one way and act differently. Our thoughts and actions must agree in purpose. Thoughts are deeds in the mental realm just as actions are deeds in the physical realm, and sooner or later, thoughts dwelled upon in the mental realm are transformed into actions, good or bad, in the physical realm.

**Evil thoughts do not always manifest as evil actions
but evil actions always begin as evil thoughts.**

Each interaction, each experience that we encounter as a life situation arises from the manner in which we responded to and reacted to previous life situations, and each experience we face offers us the opportunity to grow spiritually. Spiritual growth only occurs if the choices we make as part of the experience are acceptable to God and are made with the desire to manifest the will of God and subdue our tendency toward selfish behavior. When the mind of man misdirects spirit by entertaining and manifesting desires that are not consistent with this original impulse, it dishonors and disparages its creator and exists in a state of sin and separation from God. Every decision or response made by the mind for selfish intent or for personal gain, instead of being for the glory of God or in loving service to another person, is a lost opportunity that causes our soul to become out of tune with God.

Why can't the soul be awakened in its natural environment, the spiritual realm where it was conceived and created? What makes the earth a place where souls can recognize that they have separated themselves from God by their thought and activity in the spiritual realm? According to the readings, the causality of the physical universe as perceived by the conscious mind through time and space in combination with the soul quality of patience is the best method to reawaken the soul to God consciousness.

Timeless Eternity

> *God is not mocked, and whatsoever a soul soweth, that must it somewhere, sometime reap. For in the consciousness of eternity, time is not, neither is space. In man's consciousness there appears so much mercy, so much love, that these have been called time and space. [Edgar Cayce reading 3660-1]*

Reading 3660-1 does not say that time and space are not part of the structure of eternity, which we might take to mean the spiritual realm, but that time and space are not a part of the consciousness of eternity, presumably meaning that souls in the spiritual realm whose awareness primarily is through the subconscious mind do not have the same perception of time and space as man, whose awareness is derived through the physical senses and physical consciousness. It suggests that the concepts of time and space as experienced through physical consciousness are in some way related to or synonymous with mercy and love. This is likely a reference to the fact that the universe was conceived and created as an act of love by God whereby souls could experience a causal environment through the perception of time and space and thereby perceive and come to understand their loss of God consciousness. The quality of mercy arises because the universe is also an extension of God's grace through his willingness to provide that opportunity even though errant souls did nothing to deserve it.

Other readings describe the non-material realm occupied by the soul between earthly incarnations as a timeless realm. Selfish choices can also be made in the spiritual realm by the soul mind for reasons and gratifications that we are unable to fully comprehend from a physical perspective, but apparently it is difficult for the soul mind to understand the repercussions of these choices and activities in the spiritual realm. How the soul functions in a timeless realm is unfathomable to our finite conscious mind, just as it is impossible to imagine the structure of whatever passes for the equivalent of space in the spiritual realm and to imagine communication through the mental activity of thought.

> *For to the subconscious there is no past or future -all present. This would be well to remember in much of the information as may be given through such forces as these. [Edgar Cayce reading 136-54]*

> *Remember, in spirit there is no time but now. [Edgar Cayce reading 2560-1]*

Individuals attempting to express their near-death experiences often describe seeing themselves and others in bodies similar to their earthly bodies (Burke, 2015) and describe bright and colorful scenes that are similar to the natural world (flowers, trees, etc.) or the man-made world (houses, roads, parks, etc.). These descriptions may be accurate accounts of the spiritual world or may be carryover images in the mind that imitate the recent visual images these persons experienced during their present life. It may be that with prolonged residence in the spiritual realm these initial images fade in the same manner that childhood memories and images fade as we grow older and that the reality of the spiritual world is far different from these fleeting first impressions. Investigations of near-death experiences have not been able to definitively confirm that they represent journeys into a spiritual afterlife realm. However, some reports describe examples of cognitive perception that are difficult to explain within the framework of accepted science (Greyson, 2008 and Stevenson and Greyson, 1979).

Electromagnetic energy, the basic building material of the universe, is described in the readings as a form of spirit in motion, a physical energetic expression of spirit. Electromagnetic waves filled the universe from the moment of its creation but were only freed from incessant collisions with electrons after about 400,000 years when the sea of electrons became bound to nuclei and formed stable atoms. When these trapped waves of energy were released from confinement they raced outward with the expanding universe, never to stop unless they collided with and were absorbed by atomic matter. All electromagnetic waves that exist as remnant primordial radiation or that were later emitted from an atom when a bound electron moved from a higher-energy state to a lower-energy state or from a sun undergoing thermonuclear fusion, whether they are of radio, microwave, ultraviolet, infrared, visible light, X-ray, or gamma-ray frequencies, have a unique property; they travel through empty space (space devoid of atomic matter) at a constant speed of 299,793,458 meters per second.

The equations Isaac Newton derived to explain gravitational attraction and the acceleration of a mass induced by an applied force implied that time and space are absolute and unchangeable values from any perspective, and that the speed of communication across the universe is infinite. Special relativity and Maxwell's equations of electromagnetism showed that these were false concepts, that Newton's equations are low-velocity approximations of the true properties of space and time and that all light waves travel at a constant finite speed. If we try to apply Newton's implied infinite speed of light in the context of the equation of special relativity, time and space would indeed be absolute invariant quantities, but there also would be no mass in the universe because it would require an infinite amount of energy to create an infinitesimal amount of mass, and there would be no sense of cause and effect because all parts of the universe could communicate with each other instantly. The transformation properties of Maxwell's equations between any two frames of reference that are translated, rotated, or moving at constant velocity with respect to each other define a speed of causality, the maximum speed at which any two parts of the universe can communicate, and that anyone can detect this communication. The equations yield the rate of information flow within the universe and give its value as the speed at which light travels through space. But only massless particles such as photons, gluons, and gravitons can travel at this maximum rate of information exchange, and it is the only speed at which they can travel.

Relativity tells us that space and time are not independent properties of the universe, but that spacetime is the fundamental property. Objects with mass cannot travel at the speed of light, which limits their future spacetime coordinates to regions where their timelike motion always exceeds their spacelike motion; that is, the distance light travels in time will always be greater than the distance a mass can move in space at any speed it can acquire. This means that vast regions of spacetime are off limits from the point of view of our current moment and location. For example, we can't travel to a moon of Jupiter in time to watch an event that will happen thirty minutes from now because it takes light forty-three minutes to reach Jupiter from the earth and at best, with an infinite amount of energy at our disposal, we will arrive thirteen minutes too late. The event will happen without us being able to observe it as it happens. However, a light wave does not face the same obstacle.

It is true that light is massless and only travels at the speed of light. However, it is only true for an observer who is watching the light beam pass by. From the perspective of the electromagnetic wave, the situation is even more bizarre. Space contracts to zero for a unit of massless energy traveling at the speed of light. Electromagnetic waves are timeless units of energy and, unlike an object with mass, waves will travel zero spatial distance in a timeless interval. For all its unfathomable speed through space from our perspective, light goes nowhere in space, and it doesn't "experience" time. An electromagnetic wave is emitted in one region of the universe and absorbed on the other side without the "sense" of traveling through space or time. The readings declare that electromagnetic radiation is the physical manifestation within the universe that most closely resembles Spirit, and is the most fundamental expression of God in materiality. Electromagnetic radiation is Spirit in motion. Like Spirit, electromagnetic radiation exists in a timeless and spaceless state, but we observe it as an always-moving energy wave traveling at the fastest speed that anything in the universe can possibly travel.

Time and No Time, Space and No Space

> *The communication or the activity or the motivating force we find is three-dimensional - time, space and patience. Neither of these exists in fact, except in the concept of the individual as it may apply to time or space or patience. [Edgar Cayce reading 4035-1]*

> *This becomes hard to conceive in the finite mind; as does the finite mind fail to grasp the lack of or NO time. Yet out of Time, Space, Patience, is it possible for the consciousness of the finite to KNOW the infinite. [Edgar Cayce reading 262-115]*

> *Has He changed? Have the circumstances, the environs, the times changed? Are not Time, Space and Patience in thy consciousness a manifestation rather of His love, His patience, His longsuffering, His activities with the children of men? [Edgar Cayce reading 262-117]*

What is time? What is space? The readings state that the material universe is an environment in which the soul can experience space and time and that these experiences provide the soul with the opportunity to become cognizant of being separated from its creator, realize its loss of God awareness, and begin to understand the manner by which it can return to its original state of God consciousness. There are other readings that also clearly state that time and space are not really part of the physical structure of the universe and that there is no time and no space in the material universe. Are these apparently contradictory and different views of time and space and the nature of the universe compatible, or is this a serious discrepancy and inconsistency in the readings? How can time and space exist and not exist? Do either of these two descriptions of the nature of the universe have anything in common with the current understanding of the structure of the universe in the scientific community, or is science at odds with the readings when it comes to the reality of time and space?

The universe is not only designed as a theater in which the soul can act out its understanding of God and its relationship with fellow souls, but the activity is recorded for posterity, perhaps for review and study between earthly incarnations. These recordings are described as being written on the skein of time and space, or the skein between time and space, a type of recording device upon which all of our thoughts and actions are written, also called the Akashic records (Todeschi, 1998) or the Book of Life (Revelations 20:12, 1510-1 and 1226-1). Do these recordings involve electromagnetic radiation, which seems to reside at the boundary between the spiritual and physical realms, perhaps somehow existing between or partaking of both realms? What is a skein, meaning a tangled or complicated arrangement, in the sense of space and time? These recordings of the earthly activity of man and the soul may not be literally written upon time and space but are perhaps written by the conscious mind, which we use to construct time and space, out of the patterns of energy and matter that envelop us in the material world. Perhaps the structure of the subconscious mind stores this information as it analyses and evaluates physical events and plans the appropriate physical activity in response to those events. The method of recording described in the readings is not by visual images or written words as might be conveyed through the idea of a Book of Life or through Cayce's description of an old man who retrieved books from an archive as needed during a reading but as "active forces in the life of an entity." These records are also referred to as "destiny in the entrance of a soul into materiality" (903-23), which may indicate their use as soul memory in aligning the soul activity such that the soul is presented the appropriate opportunities to revisit some of the poor choices it made that were previously captured and stored in the records.

> *For each soul is the real life of the entity, and the records of the life are made much as the emanation of light from any source. For the light moves on in time, in space, and upon that skein between same are the records written by each soul in its activity through eternity; through its awareness, through its consciousnesses; not only in matter but in thought, in whatever realm that entity builds for itself in its experience, in its journey, in its activity. [Edgar Cayce reading 815-2]*

> *Upon time and space is written the thoughts, the deeds, the activities of an entity - as in relationships to its environs, its hereditary influence; as directed - or judgment drawn by or according to what the entity's ideal is. Hence, as it has been oft called, the record is God's book of remembrance; and each entity, each soul - as the activities of a single day of an entity in the material world - either makes same good or bad or indifferent, depending upon the entity's application of self toward that which is the ideal manner for the use of time, opportunity and the EXPRESSION of that for which each soul enters a material manifestation. [Edgar Cayce reading 1650-1]*

How scary is that! How is such a thing possible? All of our indiscretions and transgressions are recorded and, as revealed by the readings, are open to public scrutiny! That idea should lead many of us to think twice before acting, and some of us are probably wondering how we can slip into the recording studio and archives room and erase a few years of recordings, or at least create an eighteen-minute gap here and there. Apparently, the skein of space and time does not refer to the structure of spacetime, although that might seem preferable to some because it might suggest that the records would self-destruct as the end of the universe approaches. The subconscious mind apparently does more than absorb and store information; it analyses and interprets it and compares it to its innate but deeply buried understanding of its spiritual origin and destiny. The recording process appears to also involve the super conscious mind, so may be related to the conscience or inner moral sense that continually compares the thoughts and activities of the subconscious mind with the higher ideals that emanate from the Christ Spirit.

> *The body-mind and the soul are as real as any concrete object that we may discern. That activity of the soul as constructive through the mental attributes, or the superconscious mind's activity, is recorded, as are the activities of the forces in nature visible to man in his study or dealings with same. [Edgar Cayce reading 557-2]*

> *Upon the records of time and space are the records made, and thus the interpretations of same are through the record made there by the soul of the entity itself, read by those that may lay aside the consciousness and be one with such a record. [Edgar Cayce reading 557-2]*

What does science have to say about time and space, and is it compatible with the readings? What is time in the scientific world? Is it an inherent and essential component of the foundation of the universe? Is space an integral part of the physical framework of the universe that, along with time, scientists call the spacetime fabric and that in some way provides the underlying rigid framework upon which all forms of energy and atomic matter can exist? Are time and space only arbitrary and imagined quantities by which we assess and measure the interval between two events? Maybe space and time are only mathematical constructs that scientists use to help understand our surroundings and to develop mathematical relations to explain physical phenomena. Are they just useful but nonphysical concepts by which we attempt to quantify and measure the objective perceptions of our conscious mind as received and filtered through sequential sensory stimuli?

Entropy is a measure of symmetry and disorder in a system. The rapid expansion of the miniscule volume of the proto-universe by inflation formed an enormous low-entropy region containing a nearly uniform distribution of energy and fundamental particles. Although the universe was exceedingly hot, and its density and pressure were enormously high, entropy was low because the universe looked the same from any viewpoint; it was a highly ordered system. The initial low-entropy state of the universe changed into a higher-entropy state as the gravitational attraction of dark matter and atomic matter caused the nearly uniformly distributed matter to clump and collapse into stars. The transition toward a state of higher entropy in a natural closed system can be interpreted as occurring within the context of a flow of time, change in the distribution of energy and matter that proceeds toward disorder. But does the human body and the mind experience the same time we can associate with the progression of disorder in the universe? We have seen that the initial uniformity of the universe implied that time could be synchronized in the early universe and remain synchronized as spacetime expanded, allowing a universal time to be determined and giving one age to the entire universe. But this does not mean that a physical property of absolute time exists, only that the duration of space expansion can be measured. There is no way for the human mind to perceive this universal age or for the human body to sense the progression of this universal time.

The scientific definition of duration (the time difference between two events) and length (the distance between two locations in space) has evolved through the centuries as scientists seek to work out more reproducible standards that can be used by scientists across the world to make accurate and consistent scientific measurements. The standard second used to be defined as the amount of time needed for the earth to complete 1/86,400th of the time it takes to make one rotation on its axis. The standard meter used to be defined as the distance between two lines scribed on a certain platinum-iridium bar stored at the National Bureau of Standards when held at a temperature equal to the freezing point of water. The standard units of time and space are now derived from more fundamental properties of the universe, the vibration rate of a specific frequency of electromagnetic radiation and the speed of propagation of electromagnetic radiation in a vacuum, respectively. Vacuum in this context is meant in the sense of empty space or space without the presence of matter. Time is the more fundamental quantity in space-time measurements because standard time is derived from electromagnetic vibration rates, and standard length is derived from the constant speed of light and standard time.

The standard second is now defined using the frequency of microwave radiation emitted from a cesium-133 atom when an electron transitions from the higher to lower energy sublevels within a specific split ground-level energy state (hyperfine splitting). The frequency of the emitted microwave is taken to be 9,192,631,770 hertz (vibrations per second). Therefore, the period (time for one complete vibration) of the microwave is defined as 1/9,192,631,770th of a second. Everything in the universe is in a constant state of motion or vibration, and scientists have chosen one particular mode and frequency of vibration to define time. The definition of time does not generate a time label or time value to every moment of reality in the history or future of the universe; instead, time is defined by a standard interval between peaks or troughs in a particular electromagnetic wave associated with a reproducible event observed in one type of atom. The time standard is a duration or interval, not a fixed or absolute value tied to a specific moment.

The standard meter is defined using the speed at which light travels through a vacuum in a one-second duration of standard time. Light propagates through a vacuum at a constant speed of 299,792,458 meters per second independent of the location or speed of an observer relative to the source of the light. The speed of light is a fundamental and fixed constant of the universe. One meter is the distance light travels in 1/299,792,458th of a second, where a second is defined as given above. The units we use to measure distance, or interval in space, are based on an arbitrary choice of a particular frequency of microwave radiation used to define an interval of time. Like time, length is defined as a standard interval, not a fixed or absolute coordinate assigned to a specific location. The definition of length is derived from the definition of time; they aren't independent quantities. If the definition of time duration is changed, then the definition of length would also change.

So, the scientific definition of the standard unit of time doesn't give a specific time in the universe to which all other time can be referred. It only defines an interval of time to which other measured intervals of time can be compared. Is time an intrinsic property of the electromagnetic wave or are we just using an observable, periodic, and repeatable physical property of radiation to define a unit of duration? The scientific definition of the standard unit of space doesn't describe space; it defines an interval of distance as the distance a radiation wave front moves in a standard unit of time. Nothing in the definition applies a specific absolute and definitive coordinate to the spacetime fabric across which the radiation moves. The definitions of spatial length and temporal duration are based on our conscious perception of the universe. How can time and space be essential to the restoration of a soul to its maker if they aren't even real physical properties? Soul restoration is a mental and spiritual process but is facilitated by time and space as consciously perceived in a physical environment.

We physically experience spacetime.
We mentally experience space and time.

According to the readings, the properties of the physical environment in which our bodies reside and which make it amenable to healing the soul are time and space. The recovery of every rebellious soul's spiritual awareness and discovery of its heritage as a child of God is accomplished as the soul properly and successfully responds to the physical situations it encounters and the manner in which it navigates the mental and emotional entanglements of the personal relationships it develops while incarnated in a physical body. The key to a positive outcome to each life is to make choices that open the physical mind and soul mind to a perception of the oneness of all things, including the fact that each soul is a portion of God and that every soul must act in ways that honor that intimate relationship. As correct spiritually minded choices are made, especially during life's more difficult moments, the soul mind resonates with the Mind of God, and God is able to work with the soul for its spiritual growth and through the soul in a loving partnership that is helpful to other struggling souls. This process of restoration requires that the soul meet each opportunity with faith that God will use it for the spiritual benefit of the soul else its value can be lost as the soul attempts to take on the burden of the situation alone.

Ideally, the soul should always make choices that strengthen the mental and spiritual relationship between the soul and God and the soul and fellow souls. The subconscious mind, in concert with the conscious mind, too often makes poor choices during its physical incarnations in the material world. The attractions of physical pleasure, of attaining material advantage of position or power over another person, or of mental glorification of the ego or physical glorification of the body are difficult to overcome unless the soul is willing to seek divine help. The universe was designed to have the intrinsic properties that allow the soul to experience the effect of negative selfish choices that are attractive and self-gratifying to the mind but that draw the soul away from God. Our mental attitude and behavior in relation to God and neighbors can help us to better navigate these life situations and lead us to a closer association with God or further separate us from his spirit.

> *Also, in the interpretation of the universe, we find that time and space are concepts of the mental mind, as to an interpretation of or a study into the relationships with man and to the universal or God-consciousness. [Edgar Cayce reading 1747-5]*

Time and space, even if they only really exist in our consciousness, introduce the soul to causality where its decisions and actions have a visible or recognizable physical and mental effect on the body it inhabits and upon other souls that share the world within its sphere of influence. They provide an illusion of reality within the realm of the conscious mind that gives the mind the ability to identify and track the consequences of mental decisions and physical activity driven by spiritual and physical desires. The physical consequences of poor choices are often readily apparent. Our mind tells our body to hit someone in their face in response to a comment that angered us. We immediately see blood dripping from the unfortunate person's nose and maybe we feel the crunch of a fist against our own nose as that individual retaliates in response to our action. Perhaps the initial action was simple anger or perhaps the instinctive reaction of a survival response. The spiritual consequences of the altercation are somewhat different in each case, and some consequences may be more difficult to discern by the physical consciousness than others, but all are equally apparent as we watch our life unfold before us. Our behavior, rash or preplanned, is an expression of our willingness to ignore God's rules of behavior and may set in motion a need for us to be placed in a future similar situation where we can, ideally, make a better choice more consistent with the better behavior God desires from us. Our action may cause us to be on the receiving end of abuse from another individual struggling with the same problem of self-control to better understand the harm such an attitude and loss of control can inflict upon another.

One way to think of time is that it arises within the physical consciousness as it perceives the universe and attempts to order events detected by the body's sensory organs. It is an integrated aspect of the perception of the universe by the conscious mind. The mental experience of time introduces causality to the soul. It is not a salvation moment or the flash of revelation that brings us knowledge of our relationship with God. It is repeatedly meeting our shortcomings and selfishness through suffering, time after time, in situation after situation, until the soul recognizes the consequences of its words and actions and chooses alternative modes of expression that let selfishness and self-centeredness be purged from our conscious and subconscious minds. This can only happen in a causal world where we reap what we sow. A causal world is an ordered world where incorrect or inappropriate decisions and choices at one moment propagate into the future where they can be met as an opportunity for change at a later moment.

Whether or not time is an inherent and absolute physical property of the universe, time as a concept or consciousness is necessary for the soul mind to grasp and understand that God is Law and that the universe is a material expression of law. Through causality the soul can come to realize that thoughts, and the manifested words and actions that follow thoughts, always have spiritual consequences. Time is an expression of law, a medium through which law is brought into consciousness. The word light as used in the readings and in the Bible has a dual meaning. It is used to describe the electromagnetic energy that shines through the universe as visible light that our eyes can detect and also as the conscious awareness gained during an expansion of the perceptive powers of mind to receive and grasp new insights or understanding. The readings also connect light with time in the sense that conscious opportunity exists because the causal universe is perceived as time-ordered by the mind and the finite speed of electromagnetic energy is the feature of the matter universe that imparts the causality needed for the advancement of soul consciousness.

> *"Let there be light," then, was that consciousness that Time began to be a factor in the experience of those creatures that had entangled themselves in matter; and became what we know as the influences in a material plane. And the moving force and the life in each, and the activities in each are from the Spirit. [Edgar Cayce reading 262-115]*

If the conscious fabrication of time coordinates and orders our actions in a manner that lets us see the physical consequences of our behavior, then space is the stage upon which the ideals of our incarnated souls are expressed toward others. Both offer our incarnated souls the chance to recognize and respect God as Father or to ignore his presence and push him into the background of our consciousness. We can either treat our neighbors in a loving manner or as objects to control and manipulate. Our spiritual ideals, those highest aspirations toward which we strive, are manifested in the three-dimensional material world by the activity of the will on the mind. The readings frequently express the concept that the mind is the builder and the physical is the result. The most important events of our lives are usually lived within the context of a space and time environment largely determined by our previous behavior.

Space is the arena in which man uses his mind to respond to physical conditions that have crystallized around events it previously set in motion or events that will test the ability of the soul to make proper spiritual choices. Space forms the environment in which our choices are manifested and play out and sets the stage for the next act in the ongoing drama of soul growth or soul retardation. We facilitate and cause to materialize new future conditions depending on our choices and responses to each situation we face in life, no matter how trivial it may seem. These situations range from deciding to make an angry comment to the slow fast-food employee who is going to make you late for work to sabotaging a component of the machinery at the factory where you work because you didn't get the promotion and raise you expected from your boss. All of the self-centered, inappropriate, and ungodly interests we hold dear in our soul mind eventually manifest in our earthly lives through the physical mind. Selfish carnal desires depend on and are facilitated by biological sensory processes such as touch, smell, hearing, taste, and sight. Physical sensory phenomena are experienced in the spiritual environment of the soul through the physical consciousness and are interpreted by our conscious mind as pleasurable or painful experiences. Most selfish activities involve other incarnated souls with whom we interact through the physical body within the context and constraints of our physical environment.

Space is a grand stage on which all life events occur, a place in which the soul-directed conscious mind can manifest a desire to get even with another individual or disrupt the activities and financial health of a group of individuals because it feels cheated or belittled. William Shakespeare compared the world to a stage and the lives of humans to the actors in a play in his comedy *As You Like It:* "All the world's a stage, and all the men and women merely players: they have their exits and their entrances; and one man in his time plays many parts, his acts being seven ages." All the world is indeed a stage on which humans not only pass through the seven ages from infancy to dotard but also on which souls enter and exit and pass across as they consciously or unconsciously strive to fulfill their main purpose in life, to reawaken the sleeping awareness of their oneness with God. Do we really create future life situations and events in the spatial environment of the earth and then adjust the time and location of our return to earth to meet that flow of events? Does our soul wait in the wings for unrelated events on the earth to come together to bring about an environment that is a good approximation of the next set of conditions required by our soul so that it might come to terms with another of its shortcomings?

> *The eighth problem concerns the pattern made by parents at conception. Should it be said that this pattern attracts a certain soul because it approximates conditions which that soul wishes to work with? (A) It approximates conditions. It does not set. For, the individual entity or soul, given the opportunity, has its own free will to work in or out of those problems as presented by that very union. Yet the very union, of course, attracts or brings a channel or an opportunity for the expression of an individual entity. [Edgar Cayce reading 5749-14]*

When we think of time, we usually think of one event occurring before or after another event, which leads us to the idea of past, present, and future. We imagine a universe in which time moves or flows and there is a continuity of time from some starting point at the beginning of the universe to some distant and possibly never-ending future. Last week (in the past) you planted lettuce seeds in your garden. Next month (in the future) you will pick lettuce leaves and enjoy a tasty spring salad. That seems straightforward. If you don't plant the seeds, you won't be able to eat lettuce from your garden, and the first act necessarily precedes the second act. However, are "past" and "future" real concepts or just our minds trying to making sense of the world? Does everyone have the same sense of past and future or is your sense of time-ordered events different from everyone else?

> *How long before tomorrow? How long before or since yesterday? These are a matter of the mental concept. [Edgar Cayce reading 1490-4]*

There is no physical evidence that the activity of the universe follows some flow of absolute time. Most physical theories are time invariant; they work equally well in either direction of time flow. The theories do not support the notion that time is a series or sequence of reality moments projected onto the body to be read by the mind. Our psychological sense of now, the momentary present of our physical life is based on deception. Consider the relation between the conscious perception of physical phenomena and the actual physical events. Most of the information we process about the world in which we are immersed comes from the vibrations of electromagnetic waves and atoms as detected by and perceived through three of the senses (sight, hearing, and touch). The primary mode of interaction between our bodies and the physical universe for most individuals is through our eyes, which see the universe around us by detecting light waves that strike the retina at the back of the eyeball. It is estimated that about 200 million photons hit the retina at the back of the open eye every second.

Only a small interval of the known range of photon wavelengths can be recognized by the rods and cones that line the back of an eyeball, but even this limited range of perception allows us to form an amazingly complex and seemingly complete image of our environment. Visual reality, or the illusion of reality, is mainly formed by our conscious mind from the collection of visible wavelength photons that impinge on the light receptors in the eye. We tend to think that each fleeting image represents a momentary reality, or specific instance of time, within the progression of time along which the universe flows. We should understand that our eyes do not record a time-ordered series of physical events that represent a sequence of specific instances in time. The massive flux of photons that strikes our eyes at any particular moment originated from photon-emitting events that occurred at many different locations in spacetime. Light waves travel at a finite speed, but that speed is enormous relative to the speeds of other physical phenomena with which we are familiar. These waves often cover large spatial distances from various regions of the universe before we observe them. Our minds process the aggregate of sensations caused by many photons impinging on our eyes from events that have already occurred in many different locations throughout what we perceive as space and time, and combine this information with sound waves vibrating our ear drums and various tactile sensations from our skin to formulate the now instant of time we are experiencing. The mind has to interpret this vibrational energy to produce each mental image and organize it to perceive the constantly changing energy environment. But it only constructs an illusion of the actual universe of energy and matter and chooses to believe it is reality.

We make decisions and choices on the basis of past events. The physical events that our eyes respond to are not in our awareness until news of the event travels by light through space and by electron through nerve impulses to our brain and conscious mind. Our concept of now is defined by a broad spectrum of past events that depend on the distance of each event from our current location in spacetime. Information from a multitude of events reaches us simultaneously, but events that originate farther away from us in space take longer to arrive at our location and represent events farther in the "past." Every person, even those at rest relative to us, sees a different reality, a different now moment, because their different spatial locations place them at different distances and orientations from material objects within the common shared environment. This effect is in addition to the relativistic effects caused by persons moving at a constant or accelerating relative velocity with respect to each other.

> *The earth in its motion, held in space by that force of attraction, or detraction, or gravitation, or lack of gravitation in its force, so those things that do appear to have reality, and their reality to the human mind, have in reality passed into past conditions before they have reached the mind, for with the earth's laws, and its relations to other spheres, has to man become a past condition. [Edgar Cayce reading 3744-5]*

Imagine sitting on a bench in your backyard after dark. Within your field of vision a small campfire burns brightly 10 meters to your right, and the light of an aircraft warning beacon flashes red and white on a tower atop a hill about a mile off to the left. Rigel, the brightest star in the constellation Orion, has cleared the horizon and shines brightly in the sky in front of you. The campfire pops, and a spark jumps out of the fire as you gaze at Rigel, and from the corner of your eye you notice the warning light turns red. All of these signals move from the eyes and ears to the brain through neurons in which information travels at the relatively slow speed of about 100 feet per second. Your mind registers each of these events and decides they are part of the present moment that you call now. The "instant of time" you experience is your now moment, a brief sample of the visible electromagnetic radiation detected by your eyes and the atmospheric disturbance that vibrated your ear drums. Your body is only responding to variations in the frequency and magnitude of visible radiation and sound vibrations from the physical world in which your body is immersed, and that is all the information that the mind has to work with when it builds a physical reality for you. From these sensations your conscious mind creates a mental snapshot of the events occurring around you and your mind orders these now moments and interprets them as the flow of time. What does your current now moment tell you about the flow of time in the universe and your position in that stream of time? Nothing, because your body does not sense time in the universe, and there is no absolute time to sense.

Your mind assumed that the light and noise from the spark in the campfire, the change in color of the beacon light, and the light from Rigel represent simultaneous events because your body detected the vibrations from these events at the same moment. Actually, the light from the spark reached your eyes in about thirty billionths of a second, much faster than the thirty thousandths of a second it took the noise associated with the spark's creation to reach your ears. The light from the spark arrived at your eyes about a million times faster than the noise of the pop arrived at your ears. The change in the color of the warning light from white to red reached your eyes about five millionths of a second after the event occurred at the beacon. The light from the star Rigel is essentially a steady blue white, but its passage through the atmosphere has caused it to appear to the eye as if it is twinkling. There is no change in color that can be interpreted as being caused by a specific event that took place in the interior of the star, but the starlight you see has been racing across the universe for about 864 years before ending its journey at the back of your eyeball.

All of the information processed by the brain and interpreted by the mind during this now moment that the brain interprets as the present is old news as far as the universe is concerned. All of it came from events that already happened and some of it from events that happened centuries before your birth. The now moment your mind believes your body is experiencing is a personal now moment not shared by anyone or anything around you and is constructed by your mind as a composite of signals received from events that did not occur simultaneously. By constantly sampling the immediate environment near the body's sensory organs, every person compiles a series of snapshots of vibrating matter and energy previously radiated from various sources and uses that data to build a mental image of its surroundings. The image does not represent a single now moment in time associated with the sources of the information you are detecting. Instead, it is a smeared, shadowy, and illusive now moment pieced together by your mind over the fraction of a second during which your body collected the sensory data and your mind processed the data. The image the mind forms is a good approximation of the current state of vibrations associated with atomic and molecular processes occurring in nearby objects, but a poor indicator of the current state of processes occurring in more distant objects. Physical consciousness arises out the collection of still images compiled by the mind as it repeatedly samples the influx of neural data arriving at the brain much like a movie is built from a sequence of individual still frames. These images represent a composite of many ongoing physical events and have temporal meaning only as a consciously imagined now moment but in sequence do collectively capture changes in the motion and vibrational energy of the surrounding and evolving physical scene.

> *As is understood, Life - God - in its essence is Vibration, and - as the physical beings are of that atomic force, a portion of the same - the AWARENESS of same is as to how conscious that vibration may be made, even as we find in the physical body that sight, hearing, taste, speech, are but an alteration of vibration attuned to those portions in the consciousness of the physical body … . [Edgar Cayce reading 281-4]*

What if you weren't alone while sitting on the bench, and your eyes were sharing the night sights and sounds with a friend sitting next to you to your left, perhaps enjoying a pleasurable moment with a loved one? You might think you are sharing a special moment in time, but you aren't even sharing the same now moment, let alone a particular instant of absolute time in the history of the universe. At the instant of your personal now moment previously described, your friend has not heard the pop associated with the spark being thrown from the fire or seen the flash of its light because those events belong to your companion's future. Your friend has already seen the beacon change from white to red so that transition is part of his or her perceived past, but your views of the twinkling starlight are synchronized. Your friend's now moment is no more and no less real than your now moment. His or her sense of past, present, and future are just as valid as yours. Yet they are slightly different perceptions of the same physical universe that you believe you are sharing at the same time.

The perception of now for each individual is based on a different but very similar sampling of the current evolutionary state of the universe. Because of the proximity of the friend and the rapid speed at which light propagates, and to a lesser degree the speed at which sound propagates, both viewers think they are sharing and sensing the same reality, the same now moment. If the speed of light and speed of sound were much slower or the separation of the friends was much larger, the information they would receive from the universe around them, and their perceived now moment, would be more different. They would be able to mentally assess that they perceived a different view of the state of the universe but would still not be able to intuit that their personal now moments do not lie within a particular shared absolute universal time.

There are, as were set in the beginning, as far as the concern is of this physical earth plane, those rules or laws in the relative force of those that govern the earth, and the beings of the earth plane, and also that same law governs the planets, stars, constellations, groups, that that constitutes the sphere, the space, in which the planet moves. These are of the one force, and we see the manifestation of the relation of one force with another in the many various phases as is shown, for in fact that which to the human mind exists, in fact does not exist, for it has been in past before it is to the human mind in existence. [Edgar Cayce reading 3744-5]

Patience

For as Time and Space are manifestations of Divinity in the experience of individuals in material plane, the other - Patience - must be ever an attribute of an entity in making application of self and self 's virtues; overcoming faults or correcting this or that in the experience, for then ye may find these will work together for good in the experience. [Edgar Cayce reading 1510-1]

And in Patience then does man become more and more aware OF the continuity of life, of his soul being a portion of the Whole; Patience being the portion of man's sphere of activity in the finite being, as Time and Space manifest the creative and motivative force. [Edgar Cayce reading 1554-3]

So where does patience come in and what does it have to do with the physical world and soul growth? The readings declare that the exercise of active patience by incarnated souls can potentially free the soul mind from the self-inflicted illusion that it is a separate and independent being from God. In metallurgy, metallic ore is tested by heating a sample to a high temperature in a crucible. The ore is then proved by extraction of pure metal from the sample. Like a piece of metallic ore, every soul is being tested, refined, and proved in the crucible of the earth. The process of refining the soul will continue until it is proved worthy of companionship with God. But refinement is not something that God does to the incarnated soul. To be successful, the refinement process requires cooperation between the soul and God. Cooperation, as defined in the readings, is more than helping one another to achieve a common goal. It is the opening of the soul mind to receive the Christ Spirit so that the attributes of God might be made manifest in the world. This is a healing process that realigns the soul with its creator and, once again, allows the part, the smallest unit of independent spirit in the spiritual realm to fulfill its duty and destiny within God, the greater whole. Jesus could do no great healing works near his childhood home. The forces he used to bring healing were not transmitted through space from him to the ill person, but were transmitted spiritually and mentally from God through the soul mind of the ill person, thence to the body of the ill person. All healing is from God, and Jesus had to awaken the soul of the ill person to a level of faith that would permit it to receive and accept that healing energy. If a sick person that Jesus wished to heal held on to an attitude of disbelief and incredulity or if the individuals that gathered round to witness the event were doubtful of the outcome because of their conscious memories of

Jesus as a child of the village carpenter in Nazareth, it could create a mental barrier that would inhibit or block the flow of God's healing energy to the sick person's body.

> *... all healing, all resuscitating forces of a body must come from the divine from within, and to some this may come only to those who are enabled to gain such a consciousness from within that will make for CONSTRUCTIVE forces .... [Edgar Cayce reading 4757-1]*

Peter tried to give us some concept of the great patience of God (2 Peter 3:8) by stating that "one day is with the Lord as a thousand years, and a thousand years as one day." Where man is impatient and wants everything now, God has all eternity before him and is willing to patiently wait until the time is right for his plans to come to fruition. He does not want any soul to perish (2 Peter 3:9) but desires that all souls come to repentance and to recognize their shortcomings and failures to honor the purpose for which they were created. On our part, we need to repent of our sinful thoughts and actions, ask for his forgiveness, and alter our mental attitude and physical behavior to manifest and reinforce our change in consciousness. If we truly think that God has such extreme patience with us as he waits for us to return to the awareness and consciousness of our oneness with him, why do Christians in the United States, for example, believe that his patience only lasts about 78.5 years, the average lifespan for a person born in this country, at which point he is apparently willing to consign a significant portion of its population to an eternal hell? Do Christians really believe that God is that impatient?

Patience is an essential quality we must embody to overcome soul entrapment in the material world. It allows us to overlook slights and slurs from others and respond in a manner that does not entangle us in karmic repercussions. Jesus said that we are capable of doing the things that he did when he walked the earth. On the spiritual level we are like Jesus; we are created in the image of God with all of the attributes and capabilities of a soul, a spiritual being. But as long as we remain mentally attached to the material sensual world, we will not be able to recognize this relationship or actualize the mental and physical abilities that are inherent within us because of the spiritual attributes our souls possess. We can only lift ourselves out of this quagmire by following the path demonstrated, taught by, and exampled by Jesus and by seeking help and guidance from the Christ Spirit that is present within us and available to us from our higher mind. This is what Jesus asks of us. He does not want us to worship him because of what he did and endured for our sakes, but wants us to emulate his life by being of service to others as the opportunity and need arises.

A soul must be willing to cooperate with God and exercise active patience and a positive outlook because it will continue to meet its previous failures until it has no more failures to meet. The process typically takes many lifetimes. Most souls that are active on the earth today will have more lessons to learn and a long history of failures to revisit. Patience is a spiritual quality that brings the soul to a refined state. In patience the soul's activity on the earth becomes grounded in material application of spiritual qualities and attributes such as love, kindness, and forgiveness and loses the desire for selfishness and self-centered behavior. In patience and repetition we habituate our minds to constantly and consistently express the inherent moral and spiritual qualities of the soul. The standard to which our actions should be compared to determine if these attributes are being correctly applied was given in the life of Jesus the Christ, a perfect example of a life well lived as demonstrated by his unselfish motives and actions. He was the pure essence and expression of the soul attributes every person must learn to manifest daily no matter what their religious affiliation.

The readings state that from the viewpoint of a conscious mind occupying a human body and experiencing and creating within a material environment, God is time, space, and patience. What does this mean? Is it a sacrilegious comment that degrades God to physics and psychology? Is it a profound statement that identifies the material universe as an important environment created by God for the purpose of bringing enlightenment to each incarnated soul? It is actually another way to say that God is Law, and the law of God as applied to the soul consciousness will help the errant soul to realize its separation from God. It also expresses the fact that God is Love. He is willing to let the soul disappoint him many times as it works patiently to understand its spiritual errors and shortcomings. The proper response of the soul mind to opportunities it experiences in spacetime is critical to the spiritual destiny of the soul. The soul expresses itself through the activity of the physical consciousness in the causal world. Every choice made by man originates in the soul mind (our subconscious mind) and manifests in the physical world through the direction of the physical mind. These choices made in spirit not only have physical consequences but also positively or negatively affect the attunement of the soul mind to the Mind of God and reveals or clouds the soul's innate awareness of its spiritual origin.

> *Then there is no time, there is no space, when patience becomes manifested in love. [Edgar Cayce reading 3161-1]*

For conscious awareness of time and space to be effective in changing the outlook and ideals of the soul, the soul must approach life experiences with patience, an innate attribute of every soul that must be honed and developed until it becomes part of the response to every situation encountered. Souls that incarnate on the earth are not damaged beyond repair, but are usually spiritually dysfunctional to the extent that they will require many incarnations and many life experiences to be healed and restored to a state of perfect communion with God. They must meet many different life situations of their own creation in patience to successfully resolve them, and they must perse-vere until the last obstacle to full God consciousness is realized.

There is a certain caveat that is commonly used in the readings when they state that there is no space and no time.

> *... there is no time, there is no space save as the individual use or application of same may separate it into such activities*

as necessitate the meeting of conclusions or choices sought or wrought in those experiences. [Edgar Cayce reading 2072-8]

This reading suggests that time and space exist as an aspect of the physical consciousness that can be used by the soul in the process of spiritual enlightenment rather than as absolute fundamental properties of the universe. Life on the physical earth gives the soul the setting and opportunity to discover its real nature and origins. It causes the soul to see how far it has drifted from God consciousness and to become aware of its oneness with its creator. Without patience we quickly succumb to negative attitudes such as anger, frustration, hatred, and other means of personal interaction that are contrary to God's plan and purpose for us as his companions and co-creators. Patience allows us to eventually understand how our selfishness pushes us away from God and God consciousness. Through patience we learn over time to consistently make correct moral and ethical choices. Basically, we are slow learners and need the mental reinforcement acquired by meeting our spiritual flaws and imperfections again and again before our wills can harness our minds, and we can become mentally strong enough to set aside our selfish aspects completely and let our lives become like the example shown to us by Jesus. Jesus expressed the necessity for patience and resolve and defined the ultimate reward of the exercise of patience with the statement, "In patience you possess your souls" (Luke 21:19). This statement follows several verses in which Jesus describes the trials and tribulations that will be the lot of those who declare allegiance to him and follow his way.

**Time gives the conscious realization of causality,
Space gives the awareness of material activity,
Patience gives the realization of time and space illusion
and leads to the perfection of the soul.**

Patience implies consistency over the long term, so it is not sufficient to do good one day only to do bad the next. Patience is a necessary ingredient for soul growth because the successful meeting of a particular material situation facing our soul does not grant instant translation to a higher spiritual state or immediate access to heaven or the kingdom of God. We may need to repeat the same lesson many times before it becomes an integral part of our being and we can consistently make proper choices as we move through each experience. It typically takes many incarnations for a soul to come to the realization that its thoughts and actions are separating it from God, and more incarnations for the soul to consistently make the correct decisions that will cause it to regain awareness of its original relationship with God.

> *Passive patience, to be sure, has its place; but consider patience rather from the precepts of God's relationship to man: love unbounded is patience. Love manifested is patience. Endurance at times is patience, consistence ever is patience. [Edgar Cayce reading 3161-1]*

Patience may be active or passive in its application. From a spiritual perspective, patience is more than stoically putting up with the current situation. The patience that brings souls back into a loving relationship with God is not passive acceptance of events but utilizing every opportunity that is presented to the soul to actively engage with the goal of affecting positive change. At the same time the soul must understand and accept the fact that the reward of patient behavior will come about in God's time frame, not man's time frame. This is the principle that man should make the effort but that God will give the results and increase (1 Corinthians 3:7). If a person facing an angry individual responds in kind with his own retort, neither person will have an opportunity to reap a positive outcome. The inappropriate and selfish attitudes of both persons will have been in opposition to the higher spiritual nature of their souls, and neither will have manifested the godly quality of love. Both individuals will need to face similar situations in the future to give them another opportunity to make choices that will lead them to an awareness of a different and better manner of personal interaction. The spiritual law under which both are held responsible for their actions operates to ensure that future events in their lives will unfold so that they once again meet their shortcomings and will learn to make correct, God-oriented choices, before engaging their mouths.

When a person correctly responds to an angry outburst by mentally or verbally forgiving the offending person and not responding in kind, he has created a possible path to a more positive outcome. He is doing more than just defusing an ugly situation. First, he is making a response that will begin the process of removing the need for him to be involved in these kinds of confrontations in the future. As he continues to respond in a patient and loving manner in similar situations, the soul that is directing and manifesting the loving response draws closer to God, and the soul mind becomes more aware of its true relationship with God. It eventually becomes unnecessary for the body to face those types of situations because the soul has learned that spiritual lesson and the spiritual principle has become an integral part of the soul's individuality. Secondly, the person initiating the angry encounter will see a different response to his actions than he might be expecting, and that unexpected event may trigger a memory or feeling that reminds him that his actions and words are not acceptable to God. It may be enough to change his heart or at least start some change deep inside his inner being that will eventually lead to an outward display of a change of heart. This is where active patience is important. The proper response is not simply to tolerate the anger while holding back the urge to respond in kind but to engage the person with a positive loving reply or thought. Even when the response is active and positive, the desired effect upon the instigator may take minutes, days, or lifetimes, depending on the willingness of the person to relinquish the anger, open himself to God's presence and let his soul mind and his conscious mind be transformed by God.

The flow of events in the universe, the nuclear reactions in stars, the vibration of atomic motion and the absorption, emission, and propagation of electromagnetic waves and the exchange of electrons that provides energy to living matter carry on whether or not we are incarnated. When we die and reincarnate in a later era we pick up where we left off in regards to the types of situations we must face to test the degree to which we have learned or failed to learn certain spiritual lessons. We may once again meet those we have loved and treated well and those we have hated and despised or just simply treated poorly. Those who responded to our anger and bitterness with kindness and forgiveness probably will have moved on to other experiences needed for their spiritual growth, but we will find another soul who will need the opportunity to face similar shortcomings and who can work on them within the context of a relationship or acquaintance with us. We only progress spiritually when we realize the need for better responses during those situations. We are constantly and frequently given new opportunities to look deep within ourselves and determine if we are ready to let God be our guide and accept his help as we face the conditions we have created. We can choose to meet them in a more God-like manner, or we can choose to ignore his calling and continue to use our will to inflate base, selfish, and unkind desires in our minds and continue to build our material lives in the likeness of a house of cards on a sand foundation that will never stand firm.

A microscopic measurement of position will accurately locate the present position of an electron while simultaneously erasing any knowledge of the future position of the electron. The electron's future position is defined by a new set of possibilities contained within the new probabilistic wave function created by the measurement process and environment of the electron. Are choices in the spiritual-mental-physical realm of an individual analogous to measurements in the microscopic quantum realm? Is it possible that using the soul mind to make a choice is equivalent to making an observation in the quantum physical world in that it activates or brings into focus the potential histories of the soul and defines a moment on a new history path that the soul can follow? Our spiritual future may exist as a set of probabilities governed by our responses during previous opportunities we were given to discard selfishness and manifest God's love. That future is playing out within the environment in which we are currently living and includes such features as financial security, social status, and interpersonal relationships. With each choice we make, we materialize a moment of reality on one of many possible future paths, consigning it to our past history and opening a new set of possible histories arising from our response. Our spiritual environment, our inner awareness of God's presence, and our ability to receive guidance and support from him is modified and changed with every decision. Choices made in anger or with selfish intention cause us to face future experiences that place us into similar situations where we will need to face another opportunity to make more loving choices consistent with our remarkable status as children of God. Choices we make in patience that are consistent with God's will and purpose for us alter our possible futures in ways that will lead us to a closer relationship with God and a better understanding of our true nature as souls.

Christ Consciousness In Action

The word Christ comes from the Greek word *christos* meaning anointed. When non-Christians, or even perhaps many Christians, read the Bible, they may become confused about the meaning of this word in reference to Jesus because of the inconsistent use of the word throughout the New Testament. Many verses are translated and written such that the words "Jesus Christ" convey the impression that Jesus is the given name and Christ is the family name of the man Jesus (e.g. Matthew 1:1 NIV, where the genealogy of Jesus is given). Verses in several of Paul's letters contain the words "Christ Jesus," again possibly conveying the impression of given name and surname but which also may be thought of as a title of respect followed by a given name (e.g. Acts 24:24 NIV). Translations of the form "Jesus the Christ" (Matthew 16:20 KJV) give a more accurate portrayal of the meaning of the word Christ by describing Jesus as the anointed one, but what does it mean to be anointed? The standard definition of anointed is to consecrate or sanctify by smearing or rubbing with oil. In the same context, the translated form "Jesus the Messiah" (the Hebrew equivalent of the Greek *christos* which also means anointed but is usually assumed to have the added connotation of a savior or liberator) emphasizes the role of Jesus as the anticipated Messiah of the Jews, who many expected and hoped would free the Jews from the tyranny of Roman occupation and would reign as king over a restored Jewish nation.

Other verses contain the translation "Jesus who is called Christ," where Christ can be construed as a title for Jesus, like Rabbi (Matthew 27:17 NIV). Peter uses the expression "The Christ of God" (Luke 9:20 NIV) to describe Jesus in response to Jesus's inquiry, "Who do you say that I am?" This phrase implies a spiritual kinship like the anointed messenger of God similar to the familial kinship expressed in Son of God. In the Book of Romans several verses give the impression that Christ is a physical body such as where Paul speaks of Jesus as "Christ was raised from the dead, he cannot die again" (Romans 6:9 NIV) or describes the resurrection of Jesus in the words, "he that raised up Christ from the dead" (Romans 8:11 KJV). The sense that Christ is mortal is reinforced in Galatians 2:21 (NIV) where Paul makes the statement, "For if righteousness could be gained through the law, Christ died for nothing!" In Romans 8:9 (NIV), Paul's words convey a different nature of Christ by his statement, "And if anyone does not have the Spirit of Christ, he does not belong to Christ." This suggests that Christ has a deeper meaning than "anointed," has a spiritual nature, and is something whose essence everyone should aspire to cultivate and attain. This Christ is described as having a powerful effect over sin when it dwells within the individual.

Then, who was Jesus and how did he differ from Christ, or the Spirit of Christ. What do the readings mean by Christ Consciousness, and how does it relate to souls, mankind, and Jesus? The name Jesus, if used alone without the addition of the word Christ in most of the New Testament, would be much less confusing to the reader of the Bible. Jesus is the given name of the child who was born to Joseph and Mary, who many saw as being anointed by God and destined to become the savior of the Jewish people. According to the readings and the Bible, Joseph and Mary traveled to Bethlehem to be counted in the census ordered by the Roman authorities. The readings elaborate that they were in the company of his fellow carpenters and carpenter helpers and under the protection of the Essene community (5749-15 and Imes, 2021). While waiting there for the purification period to be completed, they were warned by an angel to flee to Egypt to avoid the wrath of

Herod (Matthew 2:13–23). Upon returning from Egypt, they settled in the northern town of Nazareth where Joseph pursued his occupation as a carpenter. Jesus is commonly assumed to also have worked as a carpenter, presumably as a helper under the guidance of his father. The readings verify this but indicate that this was his occupation only briefly from ages twelve to fourteen, and later, after the death of his father, from ages twenty-five to twenty-nine [1152-4]. Between these two periods of work as a carpenter he traveled to India and Persia for education, was called back to Judea at the death of Joseph and then went to Egypt to complete his education [5749-7].

The name "Jesus" describes the man who walked and talked and taught in the cities and countryside of Judea, Samaria, and Galilee in the first century, who claimed to be the Son of God and gathered a core group of apostles and other believers and followers around him. He taught that all men should love God with all their heart, soul, mind, and strength and that they should love their neighbor as themselves. The application of these two sweeping commandments separate those who aspire to seek companionship with God from those who seek to glorify self through gratification of personal desires, accumulation of material goods, or exercise of power over others. The Jewish religious authorities felt threatened by his growing base of followers, accused him and tried him in secret for blasphemy, and obtained permission from the Roman civil authorities to execute him. His death on the cross and his subsequent resurrection and ascent into heaven is seen as a salvation event in which he gave his life for us and not only released us from our sinful condition but also rescued us from the consequences of our sins. All who declare and believe that Jesus is the Son of God and savior and who strive to follow his teachings and example may receive this salvation. Strangely enough, these basic beliefs of most Christians have little or nothing to do with Christ, except that Jesus is often referred to as Jesus Christ. Followers of Jesus call themselves Christians, and most Christians study the life of Jesus through the Gospels, the first four books of the New Testament, but many only think of the word Christ as another name for Jesus or as an honorific that is used to respect the authority he received from God.

Jesus as "the Christ" in the sense of "Jesus who became the Christ," describes Jesus as the person who was anointed by God for a special mission on the earth. Jesus was spiritually anointed and consecrated by God to fulfill his mission and plan for souls to return to an eternal state of companionship with their Maker. The attunement between Jesus the Son and God the Father was firmly established and reinforced because he made life choices consistent with God's plan for him and he lived his life in a manner that fully and continuously expressed the Christ Spirit. Jesus fully expressed God's attribute of love by accepting the will of God as his true guide and by sacrificing and submitting his personal and individual will to the service of God and humanity. The pinnacle of that sacrifice occurred when Jesus willingly allowed his physical body to be crucified on the cross for the salvation of all of humankind. All of humankind, not just those we now identify as Christians from our shortsighted and self-centered perspective. The sacrifice was for every soul that was incarnate in a living physical body during the time Jesus taught, for every soul that had incarnated before Jesus was born, and for every soul that would incarnate after Jesus's death and resurrection.

According to the readings, there is more to the term "Christ" than is apparent from its use in biblical passages. The term can be viewed from three different perspectives that have more to do with a state of mind and the relationship of the soul to God and man than with the conferring of a title or authority, although that meaning is also valid. First, Christ Consciousness is the inner awareness that comes with the realization that we are souls, that our souls are a portion of God, and that we are one with God. Second, the mind of Christ is an alignment of the will and desires of the individual to the will and purpose of God, a frame of mind that all souls should seek and strive to acquire. The mind of Christ is the mental state that precedes and guides all manifestations of God's Spirit. Third, the Christ Spirit is that aspect of the Christ that can be manifested in the material world. The mind is the builder and manifests and shapes spirit to satisfy desire. The soul that holds to the mind of Christ can manifest the Christ Spirit as a form of cooperation between the Spirit of God and the soul of man. The Christ Spirit is manifested on the earth through acts of love, kindness, patience, and similar God-given soul qualities that are imbedded in the super conscious mind of the soul and that must be expressed in materiality to bring the soul into a closer association with God and to bring the kingdom of God into the earth.

Divine truth arises from Christ Consciousness, the soul's awareness of its oneness with God.

Jesus as the Christ describes the special mental and spiritual relationship of Jesus the Son with God the Father. For Jesus to have been the Christ meant that he had an abiding inner awareness of the presence of the Spirit of God and conscious knowledge of his soul's oneness with God, the state of being called Christ Consciousness in the readings (262-29). Consider the definition of the Christ Consciousness as given in reading 5749-14, "The awareness within each soul, imprinted in pattern on the mind, waiting to be awakened by will, of the soul's oneness with God." This statement expresses the true nature of all souls as portions of God and their potential to realize a restored relationship with God. The pattern of the Christ Consciousness already is present in every discarnate soul that is active in the spiritual plane and every incarnate soul that inhabits a body on the earth plane, but it is dormant until the soul or man seeks to find it and lives in a way that will reveal it. The conscious awareness of this pattern and the inner presence of God, and the ensuing ability to be in communication with the Spirit of God within our mind is dependent upon our willful decision and desire to awaken that pattern in our lives. Christ Consciousness is the ideal mental state of soul and man. It causes all thoughts and actions to be grounded in truth. It is the source of our conscience, the origin of our moral sense and higher standards of behavior and the still, small voice within that is ever-ready to guide us safely through the trials of life and bring us closer to God. It erases all doubt from the mind and exposes all decisions to the guiding influence of Spirit.

> **Jesus didn't just come into the world to
> show us he is the Son of God.
> Jesus also came into the world to show us
> we are sons and daughters of God.**

The state of Christ Consciousness is achieved through the action of the will. The deeper recesses of the mind still contain the pattern and knowledge of the soul's oneness with God but that understanding, which was in the forefront of consciousness when the soul was created, is now buried under layers of thought patterns that were formed from selfish intents and purposes as the soul sought self-gratification, separated itself from its true spiritual purpose, and lost the knowledge of its true relationship with God. When the will of the soul consistently and passionately guides the body to express love in the material world, the soul mind can once again come to the realization of this original intimate relationship and rediscover the lost (in consciousness) mental pattern that still holds the soul's understanding of its oneness with the Creator. The key to setting into motion this process of soul reawakening is by reorienting the soul mind toward the desire for renewed spiritual and mental relationship with God and by making the Christ Spirit active in personal relationships in the physical world. It is necessary that the will of each individual make daily conscious choices that bring the mind into this state of harmony with the Will and Mind of God. The will is a bridle with a sturdy bit. When the will of the soul chooses to remain aligned with God's will and keeps a tight rein on the mind to curb selfish attitudes, the mind and body can be directed and channeled into activities that glorify God and allow the soul to comprehend that it is an expression of God and that the body is only the vehicle through which the soul expresses its concept of God on the earth.

> **Christ Consciousness is the confirmation
> and culmination of faith.**

Christ Consciousness gives us the sure knowledge of the presence and activity of God within our inner being. It engenders the spiritual growth of the soul (Figure 4). It allowed Jesus to remain faithful to God during his life and to manifest his spiritual awareness in the physical world through his thoughts, words, and actions. This spiritual alignment allowed him to remain focused on his mission despite the temptations and tribulations he faced in the world. He was able to express himself as a selfless servant for the advancement of the purposes of God and show unconditional love to his fellow souls even in the face of death.

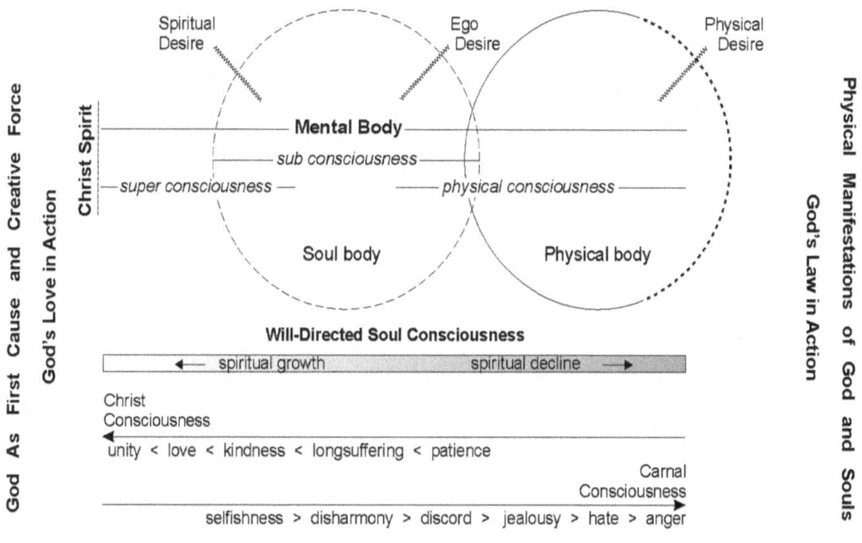

Figure 4. The mental body and the role of will in Christ Consciousness.

These manifestations of the Christ Spirit are seen in human relationships by the attitude and actions described in scripture as the fruits of the spirit (Galatians 5:22–23 NIV; love, joy, peace, patience, kindness, goodness, faithfulness, gentleness, and self-control) and all other words and actions that project love for our neighbor and reject selfish indulgence. This conscious awakening is made manifest in the physical body primarily by the influence of the mind on the pineal gland, which directs the activities of the body and can manifest the will-directed desires of the soul.

... that within same [pineal gland] that may be awakened

> *that makes for those impulses known as attributes that may
> be added to, that makes for the spiritual life first in that of
> faith as an opening, that of hope, then to virtue, then to
> knowledge, then to brotherly love, then to understanding,
> then to the Christ Consciousness, then to LOVE itself ... !*
> *[Edgar Cayce reading 294-141]*

The relationships among the man Jesus, his inner awareness of his oneness with God through Christ Consciousness, and the physical expression of this awareness as the Christ Spirit made manifest in the world allow us to see how salvation, the turning away from sin and return to God, involves more than a simple expression of belief in Jesus as savior and an expression of repentance. Jesus does not simply ask us to believe that he is the Son of God sent to deliver good news for our spiritual benefit and salvation. He requires us to open our minds to the possibility and probability that the same physical-spiritual connection that exists between him and God also exists within us. When we acknowledge that relationship we are compelled to express the Christ Spirit in every aspect of our lives. The process of our salvation begins with the faith and truth that God is present within us and his counsel and guidance is available to us. The activity of the Christ Spirit becomes apparent in our lives when we hold the mind of Christ and will nourish and strengthen our soul as we apply the message of the Christ Spirit in our dealings with others. When Jesus said, "I am the Way," he meant that his life is the ideal pattern and example for us to follow and the perfect expression of a life given to God. We must follow the path he blazed and exemplified as the Way to truly understand God and restore our soul to companionship with God.

What happened to Christ in Christian theology? Or, from the perspective of the readings, what happened to the Christ Spirit and Christ Consciousness in Christianity? Part of the problem begins with the events described in Matthew 16:16 when Peter finally understands the true nature of Jesus and the spiritual force that was his link with the divine, the source of his spiritual knowledge and the strength that would allow him to endure the cross and fulfill his mission to human-kind. Peter had a eureka moment that transcended his usual powers of deduction, a flash of spiritual lightening that illuminated his higher mind and allowed him a brief glimpse of spiritual truth well beyond his current understanding. Peter realized that Jesus is the Christ, not simply in the sense of the anointed one, but in the sense of having a shared consciousness with the Creator, a far deeper and more intimate relationship than that suggested by the title of "anointed one." This declaration that Jesus was a human who retained full consciousness of his relationship with God is not just what defines Jesus and justifies his claim to be the Way, the Truth, and the Life for the rest of humanity but it is also the foundation upon which the rest of humanity can stand firm against the onslaught of temptation faced by the soul as it attempts to navigate the material world and find its way back to God awareness.

For his great insight, later followers of Jesus would elevate Peter to a demigod status and claim that he was chosen by Jesus as the first of a long lineage of special spiritual leaders to whom the rest of mankind should show obeisance and obedience. Peter was never the strength and authority of a worldwide church established by Jesus. The Spirit of God that dwells within every man and that speaks to us as the Christ Spirit is the strength of the church, which is composed of the body of individuals who chose to acknowledge the Christ within and follow the teachings of Jesus and is not a manmade organization promoting a selected doctrine. Jesus was empowered by the Christ Spirit to represent God on the earth and was authorized to teach the proper and only approach to the salvation of the soul and the building of the kingdom of God. The church that God seeks isn't one with an authoritative fallible human being as the head of a vast organization with layer upon layer of command hierarchy who can promulgate numerous decrees and pronounce the sanction of anathema on any who dare entertain thoughts other than those ordained correct. The church that God wants is a community of believers who will seek his guidance within their heart and mind for every trial they face and for every choice they make in life and who will faithfully apply the second commandment of Jesus as the basis for all interpersonal and societal relationships. The response of Jesus to Peter's intuitive insight and fuller understanding of the real relationship between Jesus and God would be used by men more interested in attaining and wielding religious power than spreading the message of Jesus to justify taking up the mantle that only belongs to God (Ephesians 1:22–23). There is no need for selected and ordained "earthly representatives of Christ," only a need for each individual to express and manifest the Christ Spirit on the earth. The early Christian church served a purpose, but unfortunately it served itself as much or more than it served the faithful.

The universe of spacetime that we inhabit imparts a perceived causality of space and time to the human consciousness and creates a physical framework that ensures every incarnated soul will reap what it sows. As it sows the seeds of love, kindness, patience, and unity of purpose in Christ, it reaps a better understanding of its relationship with God and becomes more aware that it is a child of God. It moves closer to God in consciousness and reaps the ultimate reward of Christ Consciousness, the absolute certainty that it is a portion of God and exists in a state of eternal Oneness with its Creator. If it sows the seeds of hatred, anger, divisiveness, and disunity through selfishness and the desire for self-gratification and self-indulgence, it reaps the dubious reward of separation from God, loss of conscious awareness of God, and it obstructs the efforts of God in the form of the Christ Spirit to work with and through the soul and body for his purposes.

Ignorance of spiritual law does not allow us to circumvent or avoid the law. The law is always in effect, and we will continue to create conditions that force us to again and again face the messes we have created until we take concrete action to change our mental state and alter our physical behavior in ways that address our self-inflicted problems. Every soul is constantly and repeatedly offered the opportunity to correct the spiritual errors it has made by facing new situations similar to the ones they previously failed to handle in a spiritually responsible manner. The Akashic records or Book of Life provides causal continuity between soul incarnations by coordinating current soul activity with soul memories from past-life experiences and activity in different dimensions of consciousness. Previous poor choices that need to be addressed can be accessed and reviewed by the soul in preparation for its next incarnation. The contents of the records dictate the most suitable experiences and opportunities that are needed for soul development. The soul shapes its destiny one life at a time and one choice at a time. The spiritual heredity of the soul flows through these records just as the biological heredity of the human body flows through the DNA records of its ancestors.

> *In latent urges, two great influences are apparent. They may be compared to hereditary and environmental influences in the material plane. But these, of the real self, are the spiritual environments, the spiritual heredity. Spiritual heredity, then, is a combination of what the entity or soul has done with its opportunities for creative influence in this and all other experiences. That inherited is what the entity has made of such. [Edgar Cayce reading 2581-2]*

It is the purpose of all souls on the earth to express in personal relationships the godly attributes that were given to them when they were created. We behave in a manner consistent with the fruits of the spirit if our souls are attuned to the Mind of God and we dwell in a state of Christ Consciousness, that inner awareness of our oneness with God. The free will of the soul that expressed itself through the body of Jesus was fully in accord with the Will of God. There was no expression of selfishness in the life of Jesus because he willingly rejected material desires and temptations and maintained a conscious state of oneness with God throughout his life. Thus, the words of Jesus were consistent with the words that God wanted us to hear. The actions of Jesus were those that best presented his concept of God, the concept that God wanted manifested to humankind, and that allowed him to fulfill God's purpose for himself and God's plan for the salvation of the souls of mankind. He became the Way that all must follow because he submitted himself to God. He allowed God to work through him to bless those he touched, if they were receptive.

> *Though he were a Son, yet learned he obedience by the things which he suffered; And being made perfect, he became the author of eternal salvation unto all them that obey him. (Hebrews 5:8-9)*

Christianity places Jesus on a high and well-deserved pedestal that recognizes his role in bringing the message and example of salvation to humankind, his special status as the only begotten Son of God, and his ultimate sacrifice on the cross. But many Christians believe this elevated status puts the perfect life of Jesus beyond anything that man can approach or aspire to, let alone emulate. We can never be as good as him for we are sinners and he was not, but Hebrews 5:8–9 states that his perfection came as a result of the suffering he endured, Matthew 5:48 tells us that Jesus wants us to be perfect even as our Father in heaven and in John 10:34 Jesus reminds the Jews who are about to stone him for blasphemy that their own scriptures are correct when they make the claim that we are gods. These verses tell us that we are fully capable of being like Jesus if we are willing to walk the path that he scouted out for the rest of humanity to follow. His obedience to God in materiality was learned and reinforced through his experiences on the earth and he expects us to strive for the same perfection that he achieved. What does this tell us about our potential relationship with God? The readings reinforce the idea and ideal of human perfection and state that the life lived by Jesus is the true example of the Way that leads to soul perfection before God. And it should be understood that anyone can walk this path, whether they be Christian or non-Christian, and that calling oneself a Christian does not confer any special advantage over any other person who chooses to love God and neighbor before themselves and endeavors to walk the same path that Jesus took.

> *Hence as the Master has said, unless we become even as He, we may not in ANY wise enter in. [Edgar Cayce reading 262-114]*

There is a reason Jesus gave us the directive to be perfect (Matthew 5:48). There is no place in heaven, in the abiding presence of God, for discord, disharmony, jealousy, power struggles, or the desire to feel superior to or lord over others. The soul imperfections associated with human sin, the selfish actions taken and words spoken with a desire for gratification of ego, pleasure, or power in defiance of God, cannot be tolerated indefinitely within the body of God. Souls cannot abide perpetually in a condition of disharmony with their creator. Discord is not a natural state of the soul or the spiritual realm in which souls abide. Part of the perfection that Jesus was encouraging is freedom from the desire to let those unhealthy and ungodly attitudes be present in our thoughts and relationships. We should not expect to be suddenly transformed from selfish humans into selfless souls in the infinitesimal instant between life and death, and we should not expect that because we were good for the most part of our time on the earth that we will be transformed into a Christ-like being just because our souls left our bodies.

Upon death, our conscious mind, which derives nourishment from the use and abuse of physical desires and feeds our subconscious mind, will cease to be active because there will be no physical sensory data to process. However, the mental residue from the activity of the conscious mind in the physical universe will remain within and influence the subconscious mind after our death. In the absence of our physical consciousness, our subconscious mind will continue to be the active mental force that forms our thoughts and behavior under the direction of our will but without the need to incorporate the constant stream of information it once received through the physical consciousness and sensory network of the body. The transformation of our minds from a condition of selfish mental patterns to a spiritual state of unselfish mental patterns is not precipitated by or achieved by death. If our minds were already purely unselfish and acceptable to God, that perfect mental pattern would already have been demonstrated in the outward expressions of our physical lives, just as it was in the life of Jesus. He was truly Christ Consciousness in action and asks that we also live our lives guided by the mind of Christ so that we fulfill our true purpose in life through the perfect manifestation of the Christ Spirit.

The Opportunity Of A Lifetime

For, as ye do it, as ye treat, as ye think of each individual with whom ye come in contact along life's way, that entity - be it one physically attractive or one that casts off influences that are repulsive - is an OPPORTUNITY for YOU, as an individual, to glorify God, and to be aware of the Christ-Consciousness enabling you to meet every problem. [Edgar Cayce reading 2072-10]

What is the greatest opportunity life has to offer? A chance to get a good education which can lead to better job, higher wages, and the ability to better feed yourself and your family and buy better clothes? A second chance when a long struggle with a diagnosed terminal cancer results in remission of the disease and years are added to your life? What about the opportunity to express your thoughts and feelings freely and openly without fear of retribution from an authoritarian regime led by a ruthless dictator bent on subjecting the will of the populace to his control? These are good things to desire and appreciate, but they miss the point and purpose of life. The highest aspirations in life should not be just physical existence under circumstances that can reward an individual with a better education, a healthier body, or freedom of expression, although these are important goals in life. Of the many opportunities offered to us every day, the most important opportunity granted to us is the opportunity of life itself, which permits the soul to rediscover companionship with God.

Life is not only full of opportunities to express love of God and neighbor; life is a God-given opportunity for the soul to restore its relationship with God. Life itself is an opportunity, an opportunity for soul growth, an opportunity for the soul to recognize the damaging effects of self-centered desires and decisions that push the soul farther away from God, and an opportunity for the soul mind to restore its lost God consciousness. Souls that fail to understand and grasp this amazing and unmerited opportunity risk loss of their self-awareness and independent existence and most importantly lose an opportunity for eternal companionship with the source of all that exists, the Creator of life itself. All it takes is faith that God exists, a willingness to be receptive to his presence, and a desire to live according to the principles he established; love of God, love of neighbor, and a willingness to set selfishness aside for the benefit of others.

Science is the study of the myriad ways in which matter and energy interact and combine and the quest for the precise mathematical laws that govern the physical universe. As scientists continue to develop more sophisticated electronic technology and devise better instruments with which to probe the structure of atomic matter and the origins of the universe, these observations bring about discoveries that open new ways of looking and thinking about energy and matter. But where is the evidence that man is using his opportunities and experiences on the earth to gain spiritual wisdom along with this vast wealth of scientific knowledge? The hope for a better world through science will be elusive no matter how many scientific advances are made and how much material comfort is attained if science is allowed to draw the soul consciousness away from God. The hope for a better world through religion will not materialize as long as religious leaders use their position to deceive sincere followers for personal gain or ecclesiastical power, and congregations follow the words of their religious leaders instead of the words of Jesus as embodied in his two commandments. Religious institutions are supposed to be beacons of hope, but how do their embrace of bigotry and pseudoscience, their unbridled yearning for political power, and the willingness of their members to spread falsehoods and lies through the promotion of conspiracy theories, inspire hope? The real hope of humankind is that individual souls will take the opportunity of life to learn that true personal satisfaction and communal harmony comes from the spiritual advancement achieved by setting the commandments of Jesus as an ideal to be faithfully followed and applied in every life situation.

Scientific studies are opportunities to explore the creative genius of God. The world around us is an amalgam of pure electromagnetic energy moving at the speed of light and packets of concentrated energy compressed into a handful of basic particles and force carriers. All of it is in constant motion, the vibratory motion of electromagnetic energy as it propagates through the universe; the vibration of atoms in gases, liquids, and solids; the global expansion and contraction of gas and dust particles spread across spacetime in interstellar space; and the rotational velocity of moons, planets, stars, and galaxies caught up in a curved spacetime of their own creation. All of these forms of motion are an expression of energy acquired by the wave or matter object. We have the opportunity through science to discover the laws that define the mechanism and context within which this energy and matter can interact, whether it be the physical laws that control the formation of the stars and galaxies, the geologic laws that determine the formation and shape of the continents and oceans, the chemical laws that bring elements together to form higher-order compounds, or the biological laws that dictate how molecules, DNA, amino acids, and proteins cooperate to help bring life to inanimate matter.

Science may fail to recognize the hand of God at work in the wonders of nature, but it reveals the processes and laws God put into effect when he set his creation in motion. Scientific knowledge is not wrong because it does not conform to a particular literal interpretation of the Bible. Religious pseudoscientists use an arbitrary and narrow subset of physical observations and pseudoscientific conjecture to bolster their claims that science is wrong and their literal interpretation of biblical passages about the origins of the universe, life, and humanity is the real scientific truth. They ignore and disparage conflicting scientific research out of fear that scientific knowledge will somehow corrupt the faithful and replace blind devotion to their narrow interpretation of scripture with doubt and uncertainty. These actions are indicative of a weakness in faith, not a failure of science to discover the truth. Science seeks truth by observation, reasoning, theorizing, testing, and retesting. Scientific theories can be wrong, especially when first proposed, but application of the scientific method eventually corrects misconceptions and removes errors from theories and hypotheses. We can use this material truth to explore physical manifestations of God and to improve the living conditions of mankind, but we must be careful that the brightness of our intellectual brilliance doesn't blind us to spiritual truth.

The Lord God is One, and all that exists is an expression of his creativity. God is the Creative Force that brought into existence all that has ever existed in non-material or material form. God expresses himself through his creations, such as the universe of energy and matter and vibration and motion, and the animated world of living matter from the most primitive cell to the complex human body. His greatest creations are the souls that inhabit the spiritual realm and that incarnate into human bodies. These souls have caused him, and still cause him, enormous suffering, yet he continually offers them the opportunity to recognize their failure to remain true to their purpose and to live up to the spiritual and moral standards that will allow them to be reunited with him in love and companionship. God is all that is good because that is his nature, but souls, portions of God, choose to use their independent mind and will to manifest his Spirit in degraded and selfish forms that create evil, disharmony, and discord in his realm.

Unity of purpose within the Oneness of God is the only force that can unify humanity and bring peace and harmony to societies and the family of nations. God as One allows us the opportunity to fulfill the biblical scripture confirmed by Jesus that states that "ye are gods." Our souls are portions of God, and we are gods in the making, not in the sense of power and might and ability to lord over other souls but in the sense of creative beings belonging to the unity that is God and sharing in his creative activities. Our ability to reclaim our heritage as gods working in harmony with the Almighty Creative Force is predicated on our willingness and desire to reawaken the lost mental awareness of our true relationship with God and enter into a restored friendship and fellowship with him. He offers that opportunity to each of us with each life we live.

Will man, on a community or national level, ever come to realize that all souls, and therefore all humans, are one in God, and that the earth is a place of opportunity for soul growth, a physical realm with a spiritual purpose, not a rather large playground or a place to acquire desirable material goods? Will we squander another God-given opportunity to make good use of our life on the earth before we lose it to old age and death? Will man ever come to a collective understanding that we don't just pass into oblivion when the body dies? Are there any real signs of spiritual progress in the world? Is there a long-term upward increase in spiritual awareness hidden within the sweep of history and its repetitive wars, famines, and natural disasters that occur all too frequently? Will man ever learn to truly love his neighbor? The use of torture and other forms of brutal punishment and coercion has been abandoned by most, but certainly not all, societies, but man continues to devise ever better military hardware to make the art of war even more terrifying and too often is more than willing to use it to gain advantage over his neighbor.

God is Spirit, an etheric substance; Mind, the force he uses to direct and mold Spirit; and Will, the intelligence he uses to guide Mind to the completion and fulfillment of his purposes. Spirit is the foundation and substance of the material universe, all that we see, hear, touch, taste, and smell. Spirit is molded by the Creative Force that is God to form spiritual and physical realms. The physical laws flow out of Mind and Spirit and constrain the motion and interaction of energy and matter. Souls, as spirit endowed with will and mind in the likeness of God, are given the opportunity to create, to manifest in the spiritual and physical realms in ways that can bring them into a closer mental and spiritual association with their creator. When the soul qualities that make us more like God are manifested in the physical world, our awareness expands beyond the physical consciousness, and we remember our lost heritage as children of God. The spirit is willing but the flesh is weak. Meaningful change in the societies of the earth will only come when individuals choose to use every opportunity in life to rise above the material attractions of the world and manifest good toward others by suppressing the desire to seek gratification, wealth, and power for ourselves at the expense of others.

God is Love, the essence of pure love, the standard against which all love must be measured and the most perfect expression of love in the spiritual and material realms. His love is bountiful, limitless, and unconditional. Even God can become lonely, a condition that moved him to create souls as companions who could share in the joy and excitement of his creative activities, as companions who could comprehend the power of love and joyfully receive the love he desires to shower upon them, and as companions that have the ability to reciprocate his perfect love by expressing heartfelt love toward him. God created souls in his image and granted each of them an individual and unconstrained mind directed by an independent will. The word "image" has nothing to do with physical features or the spiritual equivalent of a bodily form but refers to the ability to express mental and emotional qualities such as love, patience, longsuffering, kindness, and mercy. God is not a person, but through the power of love, our souls can choose to make our relationship with him as personal as we desire. God is the personable approachable Father who loves his children and desires them to participate as companions and partners in his creative exploits. God as Love gave us the opportunity to have life, not only life as a biological human body, but life as an eternal spiritual being capable of enormous creativity with the ability to express divine love. God's love is abundant and sufficient for every created soul, and we have the opportunity to bask in it for eternity.

Souls are a portion of God and are endowed with godly qualities and the potential to be fully aware of their oneness with God yet having the freedom and ability to pursue their own goals and desires. From the beginning of its existence, each soul has had the opportunity to use its independent will to fulfill its purpose as a companion to a loving God or to misuse its will to descend into a self-created hell leading to mental separation from God. The transgressions perpetrated by souls against God, the fall from grace, and the loss of awareness of our honored position as soul children of God started as spiritual rebellion against God's purpose and will through selfish thoughts and actions. All sin is a willful expression of soul rebellion against God, an activity that God could not prevent without destroying the independence and free will he had granted each soul. According to the readings, one of the more important spiritual goals that a person can pursue is to use the opportunity of life to become attuned to the Mind of God, a procedure that allows the soul to commune with God and anchors the soul mind in spiritual truth.

God as Law defined the rules and boundaries within which souls must operate. He established laws for the preservation of errant souls out of the mercy and grace that is an innate characteristic of his being. Unbridled and unprincipled creative power can cause havoc in the spiritual and physical realms. God is the impersonal rule of law that sets metes and bounds on the activity of spiritual beings, and causes each soul to come face-to-face with its shortcomings and selfish behavior. Because he is impersonal law, he is also impartial law, is no respecter of persons, and is not impressed by the personalities by which men mask their true identities. God as Law offers us the opportunity to meet our many failures as souls and as soul-animated human beings. He does not infringe on our free will. He gives us the opportunity to make our own choices in response to each situation we face but defines the physical and spiritual consequences for each thought, action, and word. Our trials and tribulations are of our own creation. We are responsible for our own personal heaven or hell, and we are responsible for our salvation by the response we give to his offer of grace, mercy, and opportunity. He is also our conscience that intuitively tells us when we have erred and need to mend our ways and that allows us to recognize and empathize with the goodness, decency, and kindness we see in others.

God as Law established the rules that give the souls of men the opportunity to realize that their selfish thoughts and actions have caused them to lose conscious awareness of their creator and the knowledge of their true relationship with him. The universe exists for the benefit of souls and not the benefit of the human bodies they occupy, souls that wandered too far from their creator and can no longer easily communicate with him. The earth is a spiritual reeducation center and rehabilitation facility for souls. The law is visible in all that comes to us as bad experiences, trials, temptations, and fear, not because he wishes to harm us or wants to watch us harm ourselves and others but because he has extended his grace and mercy to us and has offered us an opportunity to extract ourselves from our terrible predicament. We create the messes in which we are immersed and that so often seem to overwhelm us. We can decide that enough is enough and ask God to show us the way that leads us out of our self-created pain and suffering. It is up to us through faith to begin the long walk that leads us back to companionship and communion with our Father. Our success depends upon our willingness to ask God to lead the way, to listen to the Christ Spirit within us for guidance, and to keep the faith that everything we see as an obstacle will be turned into a stepping stone on the path from repentance to restoration if we approach it properly.

God designed the universe so that we, as souls, can perceive the spiritual and physical damage we inflict when we become self-centered, lose sight of God, and fail to love our neighbor as ourselves. The universe of matter is the arena where we use our minds to create relationships that are based on the constructive expression of Godlike consciousness or the destructive expression of our self-centered desire for self-gratification and self-glorification. It was conceived and set on the course of stellar and planetary evolution with infinite patience and the knowledge that the soul would have the best opportunity of realizing its separation from God by facing the consequences of its rebellious actions within a time-ordered causal environment. The finite speed of light in materiality and the illusion of time in the conscious mind instill a sense of causality in the soul mind. Causality fulfills the requirement that souls must reap what they sow. Causality confronts the soul with the effects of its thoughts and its manifested actions and provides an opportunity for the soul to understand the consequences of its behavior in a way not possible in a timeless spiritual realm. Causality brings spiritual error into sharp focus, pain and suffering to the body, and mental anguish to the soul mind. This suffering is not imposed by God upon the soul, but is a reflection of the fact that the soul is no longer attuned to its Creator.

The senseless cycles of hate, division, racism, war, and other evil perpetrated and manifested by souls will end only when souls, one by one, realize that their self-induced lack of God awareness and separation from God is the root cause of all of it and make the conscious decision to change. The only way to rediscover God is to make every thought, action, and word conform to the commandments of Jesus to love God with all your heart, mind, body, and soul and to love your neighbor as yourself. This is the Law of God as applied to souls incarnated in the physical world. This is the Way taught by Jesus, and there is no other way devised by any religion that is more effective in bringing the soul to salvation and oneness with God. Love of neighbor is accomplished through the application of the fruits of the spirit (kindness, love, patience, long-suffering, etc.) in interpersonal relationships. Salvation is a process that occurs in the causal world of space and time when souls meet each challenge with patience and let life unfold one opportunity at a time.

God is the source of Life that fills the earth with biological organisms. The biological diversity of the earth is part of an ongoing natural evolutionary progression that glorifies God with its activity and that builds the environment needed to meet the nutritional requirements of the human body, the perfect body he designed for the temporary use of the soul as it embarks on the long journey of spiritual rehabilitation. God as Life allows souls the opportunity to enter human bodies and thereby come face-to-face with their shortcomings and failure to remain faithful to him in spirit. The human body is a gift from God and should be treated with the respect and honor that it deserves, with the understanding that it is subservient to the incarnated soul. The human body experiences the physical world in a variety of ways, and the mind constructs a reality that is an illusion of the true nature of the physical world of energy and matter but which is the conscious perception needed to introduce the subconscious mind to causality and responsibility for its thoughts and actions.

> *Then those that had made selfish movements moved into that which was and is OPPORTUNITY, and there came life into same. [Edgar Cayce reading 262-114]*

The human body is the highest life form on the earth and is patterned on the species of the animal kingdom that had attained the highest natural evolutionary advancement at the time of its creation. It was purposely designed to facilitate soul growth. The surface of the human body is in near-constant contact with electromagnetic radiation and some form of atomic matter such as gases, solids, and liquids. The neural cells of the brain organize the cacophony of information received from the senses by assigning the incoming data to temporary storage in specific brain cells or groups of cells according to the origin of the information and its intensity or frequency. The conscious mind sorts, orders, and interprets the data to form a mental image of its current environment and the soul mind compares the image with previously received images to form a series of now moments that it interprets as occurring in time. The soul becomes aware of the location and movement of physical objects that make up the broader environment in the vicinity of the body by inference from changes in detected stimuli.

By directing the brain to send out electrical impulses through outgoing neurons, the mind is able to activate certain muscles of the body to influence and alter its local environment by adding energy that can impact or change the motion of physical objects. Our direct influence on the world is largely limited to applying force through the use of muscles, but that relatively small force can have tremendous emotional power when it is used to move a small pen that directs a tiny flow of ink to form words and paragraphs to express ideas and when it is used to vibrate air molecules in the throat to direct words of love or hate toward another body and mind. This complex body is being directed by an even more complex soul through the activity of the subconscious mind. The soul manifests its character, its love of God or lack thereof, as it guides the body into physical activity. It can observe the effect of this physical activity upon God's creations, the inanimate matter and living creatures of the natural world and the human bodies being guided by other souls. Through physical activity and sensory stimuli the soul affects the physical, emotional, and mental well-being of other souls and recognizes the damage it is doing or the good it is doing. The incarnated soul must acquire a conscience to grow spiritually. It can only recognize whether it is living up to the high standards set before it by God when it was created if and when it is willing to open itself to the higher spiritual consciousness that compares its physical activity to the moral and spiritual standards set by God. That is the whole purpose of the physical universe and soul incarnation into human bodies. The real question we must ask is this: will the soul make use of this unique opportunity?

> *For He hath not willed that any soul should perish, but hath*
> *with every temptation, every fault, prepared a way, a manner,*
> *an opportunity for the entity to become as one with Him.*
> *[Edgar Cayce reading 257-201]*

God is Truth, the truth that transcends the accumulated knowledge, understanding, and wisdom of man who relies on inductive and deductive reasoning, and occasionally intuition, to acquire knowledge. He is the truth that sweeps away ignorance, hatred, bigotry, duplicity, anger, and all other misapplications of spirit. He is the truth that dispels the lies that men spin for their own selfish purposes, the reality behind the illusion of time and space. The truth about God is not found in the literal interpretation of the Bible or in religious doctrine. Christians routinely read the Bible and come to completely different concepts of truth. Seeking to impose ones perceived religious truth upon another person may advance religion, or the cause of a particular denomination, but does little or nothing to spiritually awaken the soul. Religious institutions should not be in the business of enforcing certain beliefs through physical or mental control such as ridicule, intimidation, or punishment. These activities do not bring souls to a better understanding of God.

Jesus did not come to the earth to preach the correct theology for humankind to believe. Jesus did not condemn or ostracize those who did not follow him. He came to present the truth, to educate the masses about a way of life that would lead them away from materialism and selfishness and put their souls on the path to restoration and realignment with God and neighbor. Spiritual truth is verifiable. The truth of religious doctrine can be verified or rejected by its observed impact on individuals and society in relation to the two commandments of Jesus. Denying science to protect religion does not serve God or bolster the truth of the reality that God exists. God as Truth gives man the opportunity to escape the self-imposed ignorance of his true nature as a soul when he faithfully seeks knowledge of God through communion with the Christ Spirit.

> *Man's discerning of that he would worship, then, is within self; how that he, the individual entity, makes manifest in his dealings daily with his fellow men that God is, and that the individual entity is - in body, in mind, in soul - a witness of such; and thus he loveth, he treateth his brother as himself. [Edgar Cayce reading 2879-1]*

Truth received from the physical world through the words and actions of others should always be compared with truth received from the inner spiritual world by opening one's consciousness to the presence and voice of God. No one has perfect access to the truth unless they have perfect attunement to God. Truth by committee, whether from a religious or political organization, is by definition not the absolute truth but only a series of compromises by sometimes well-meaning people and sometimes deceitful individuals who are more interested in power or attention than discovering and disseminating the truth. Early Christian church leaders often dealt in power more than truth and often used truth formulated by committee to hold power over the uneducated masses that could not and were not allowed to read the accumulated spiritual writings that circulated among the educated in the first few centuries after the death of Jesus (Martin, 1981).

There is a caveat to the maxim that each individual should seek truth within himself or herself. The very fact that we are here on the earth as incarnated souls is evidence that we don't have a clear understanding of the truth. We have so abused and misused our mind's creative ability in the pursuit of selfish interests that we have built a mental barrier between ourselves and God. When we do seek the truth within, we often come face-to-face with a mental fog that reflects our own biases and misconceptions. The fog in our mind reflects our own desires and false ideas and prevents our mind from penetrating to the God consciousness within us and prevents us from seeing the truth. This fog is constructed layer upon layer with every word we utter and action we take that seeks to glorify ourselves or give us advantage over another. So how do we pierce the mental fog that prevents us from mentally penetrating beyond our physical conscious mind and into our super conscious mind where God is waiting to meet us and speak with us?

We must actively seek God within ourselves, not passively wait for him to take the first step. He has already taken several steps. He created this universe of spiritual opportunity where we can live in human form in a realm of causality and where we can recognize and remove the mental fog that impedes our ability to commune with him. He sent his firstborn soul to the earth as the leader of a large group of dedicated followers and coworkers to spread the message worldwide that God wants all souls to return to communion with him. He has repeatedly sent prophets throughout human history to encourage humanity to turn back to him. He sent his firstborn soul in the form of his human Son to demonstrate by his life and death that there is more to life than living and more to death than dying. Humanity can seek God through prayer, contemplation, meditation, spiritual study, and the setting of high spiritual and moral ideals that we habitually strive to meet. We must put aside mental desires that conflict with the commandments of Jesus, and we must strive to let God be the light that guides every step we take back to him. As we persevere with faith that God will guide us safely through the fog, the fog will lift, and we will see the glory that we left behind so long ago illuminated before us and realize the truth of our oneness with him.

We must also put into practice the commandment of Jesus to love thy neighbor as thyself if we are to find God. This is not a feel-good activity or a way to gain favor with God that adds more weight to the good side of the good-bad scale so that we might tip the scale enough to get into heaven. This is an activity that literally alters the mind of those that practice it, helps them to perceive the truth of God through service to others who are in need, and reveals our true relationship with him. Service is an opportunity to exercise active patience that opens our mind to the love that is God and is a soul-altering activity that encourages us to seek a deeper meaning in life than the pursuit of material pleasure and power. It lets us look beyond our narrow personal interests and changes us in a way that helps lift the mental fog and allows God to work through us for the benefit of others. It shows us our true relationship with our fellow souls who are also portions of God and are lost like us and need a helping hand to get motivated to take the first step that will release them from the mire of material-oriented selfishness and reveal the mind-altering consciousness of God's presence. Truth, real truth, is revealed to us from within as we act in ways that cause our soul mind to rely less on the physical mind for its interpretation of reality and to rely more on the Christ Spirit for our mental and emotional strength and spiritual guidance from our Creator.

The readings are soul oriented and are human oriented only to the extent that every human is animated by a created soul and is an outlet for that soul to express itself in a material setting. The readings are also mind oriented because every soul uses its mind to build the human life it currently experiences, and life changes can only happen when they are preceded by changes in the mind as directed by the will of the individual soul. Scripture and the Search for God books are only of value to the extent that they are applied in our everyday lives. The readings declare that the only source of truth is God and that we can find and recognize that truth only to the extent that we set aside our selfishness and let God work through us. There are no manuals of doctrine or creed that must be accepted, just a desire to seek God and a willingness to set aside our egos and preconceived ideas and let him lead. The readings are God centered, Jesus centered, and Christ centered, not in a dogmatic way or a way that excludes non-Christians but in a way that weaves the soul, Christ Consciousness, and the Father into a glorious intimate relationship that can be ours even though we have caused our original relationship with God to become strained or severed. This truth of our kinship with God is ours to rediscover whenever we desire to reestablish that relationship and are willing to meet each opportunity in life with choices that are consistent with that desire.

God as the Christ Spirit stands ready to work with our soul to bring the knowledge of his presence within our soul mind. The Christ Spirit is ever-ready to manifest on the earth, and its activity in our lives constitutes the greatest spiritual opportunity given to man. The Christ Spirit is a force, a force that is of God and sent by God, an active force for good that can strengthen man in his trials and temptations on the earth and allow him to have a peek at God. Far too often, our selfish desires interfere with the ability of the Christ Spirit to manifest through us. Our selfish thoughts and actions block the flow of spirit from God through the soul into the material world. Too often we manifest our personal selfish desires instead of the higher ideals that the gracious loving Spirit of God offers us. Only when we are willing to turn from this love affair with self-absorption, self-centeredness, and self-gratification will we begin to lead lives that will end the cycles of abuse and suffering that frequently sweep through this world. To accomplish this task we must dampen and eliminate carnal desires that draw us away from the knowledge of his presence. We need to ask God to remove from our will and mind the desire for selfish power and the willingness to trample on others as we seek to be the dominant force in our family, community, or society. We need to ask him to eliminate from our consciousness the lust for wealth that drives us to seek satisfaction, contentment, and pleasure from material acquisitions to the extent that they detract us from seeking the presence of God in our minds and hearts and cause us to establish a mental barrier that separates us from God. We need to be receptive to the many opportunities that God provides us with during our lives. Each opportunity allows us to work with him to further his plans and purposes on this earth and to restore all souls to communion with him.

Although our souls were created by God with the freedom to express their independence through the action of free will, our souls are still intimately connected with God. The awareness of our soul's oneness with God is the culmination of a process of soul growth that leads to Christ Consciousness. Every soul is a portion of God that he set apart from himself so that we might have our own existence. To be set apart means to be independent in will and mind but does not have to mean to be separated from him in consciousness. We can become consciously aware that, as souls, we are privileged to have a spark of God within us, that we are an extension of God and are bound to him no matter how far we wander from him. Through the development of Christ Consciousness we come to understand this intimate relationship and our oneness with God and can experience this knowledge as an inner certainty rather than as an intellectual exercise. We can come to know God with the same loving attitude with which he knows us. As our minds become aware that we are extensions of God, are One with God, we have the opportunity to become channels through which the Christ Spirit, the aspect of God that interacts with the soul, can become active in the world and manifest the purposes of God.

When the mind is allowed to run free without proper guidance from the will, and when the individual will allows the mind to act on desires that are selfish, the mind loses all recognition of God. To realize God consciousness, or Christ Consciousness, the will must make the decision to conform to the will of God and actively work at guiding the mind away from the enjoyable but hazardous attractions of the physical world that deceive it and draw it away from its true purpose. An unfettered mind enjoys attachment to the physical pleasures of self-gratification and wealth, the ego trip of power, and the ability to lord over its fellow souls. All of these material attractions can seduce the mind and cause it to wander far from its original attunement with God and far from the knowledge of the soul's true purpose. The will is able to constrain the activity of the mind and demand that the mind turn away from desires that are selfish and self-gratifying and turn toward its creator. The will is able to suppress these ungodly desires, to reject them as they form in the mind, or to redirect them into channels of thought that honor God. The mind enjoys experiencing the sensory phenomena that flow to it from the physical body without consideration of their potential damaging effect to the mind and soul. The will, when it is in attunement with the will of God, well understands how damaging this attraction to the physical world can be. The will of the soul is capable of keeping the mind from becoming addicted to carnally desirable but spiritually undesirable sensual attractions. The mind, by the mere fact that the physical consciousness is connected to the body, will experience these sensations, but with the proper guidance from the will, the mind does not continually seek these experiences to the detriment of the body and the soul.

When the mind is harnessed by the will to manifest love and

goodness instead of evil, the mental barriers that have been formed by inappropriate mental and physical activity begin to break down, and the soul comes to realize that it is dependent upon God, not independent from him. As the mind and will choose to express the godly attributes inherent in the soul as part of its oneness with God, the mind once again begins to realize and understand its proper role as the builder of loving relationships. When the force that is the Christ Spirit is allowed to manifest in the physical world as loving interpersonal relationships, the soul creates harmony in the world instead of discord and chaos. At first this manifestation of goodness requires a conscious effort and the frequent redirecting of the mind by the will until the new way of thinking becomes habitual and natural to the mind.

The mental connection from God to the body is from the super conscious mind, through the subconscious mind, and to the physical conscious mind. When this collective mental body is aligned and attuned with the Mind of God, it forms an open channel or conduit that allows the free flow of the Christ Spirit from God into our lives to aid us in our struggles, trials, and temptations. When a soul is guided by the Christ Spirit, the soul is spiritually uplifted and spreads hope to fellow incarnated souls who are struggling to rediscover their true nature and their relationship with God. Jesus was fully attuned to God and received support and assurance through the activity of the Christ Spirit, which gave him the spiritual and physical strength to complete his mission. This is how the kingdom of heaven will be fulfilled on the earth. Heaven will be brought to the earth by the actions of man, the desire of man to do the will of God, and the activity of God working through man. It is the opportunity and responsibility of every individual to actively help bring the kingdom of heaven to the earth now through their thoughts and actions with respect to their fellow man.

What do we expect and desire out of life? Do we think our government owes us steady employment, a comfortable living or at least a certain basic standard of living and a safe place in which to pursue our concept of happiness in whatever form that may take? Does a rich country like America owe its citizens more than a poor nation in Africa or an overcrowded Asian country? According to the readings, there is only one thing of importance that this world and the various nations and societies that have arisen in it can give us that really matters, and that is the appropriate setting and opportunity for our soul to express its concept of God as we interact with our fellow human beings. The entire purpose for the creation of this universe, the earth, and all of the biological life that it contains is to provide an opportunity for soul growth through causal activity. Nations and societies are a reflection of how the individuals that make up their populations have expressed their concept of God during their previous and current lives. The many wars, famines, and atrocities that plague the world will only come to an end as citizen souls turn to God for spiritual strength and sustenance instead of seeking guidance from powerful political figures and listening to their empty promises and rhetoric or attempts to promote their personal self-interests. The world we live in owes us nothing beyond this God-conceived opportunity. We set the parameters that define the current and future opportunities we personally encounter, the crosses we must bear, and even the future of our nation by the manner in which we have used or misused our soul attributes of love, kindness, and patience in our interactions with our fellow citizens.

Human society offers the soul the opportunity to express its love of God and neighbor. Citizens of the world raise the spiritual decency, virtue, and morality as well as the material standard of their societies and nations when they put forth the best concept of God that they possess within their hearts and minds. We should always strive to express our highest ideals in the best manner we can, whether that means being polite to the slow waitress even though we are anxious that we might be late getting back to work or graciously accepting the ticket when we are caught speeding while trying to make up those minutes on the way back to work. Our attitudes and behavior within society are important because we project our understanding of God into society, and society reflects its collective understanding of God back to us. Only when we learn how to relate to and respond to our neighbor in the manner taught and demonstrated by Jesus are we able to grow spiritually, uplift society, and bring peace and harmony into the world. This is what Jesus wants us to do and why he stressed again and again the importance of applying his teachings in interpersonal relationships.

As we learn to love God and willingly love and serve our neighbor, we become more Christ-like and more aware that we are souls, children of God, and portions of God and that we have the capacity and potential to truly become companions of the Almighty. Jesus told us to be perfect even as God in heaven, and he didn't intend that we should settle for anything less. As we become better sons, daughters, brothers, sisters, friends, and neighbors, we will transform the societies in which we live. We may not become wealthy or powerful religious figures or important politicians, but we will become moral and spiritual lights in whatever society we reside, and the entire society will be a better place in which to live. As we cooperate in a loving manner with each other and with members of other societies and nations, peace on earth will become the norm instead of the exception.

> *As the Spirit of God once moved to bring peace and harmony out of chaos, so must the Spirit move over the earth and magnify itself in the hearts, minds and souls of men to bring peace, harmony and understanding, that they may dwell together in a way that will bring that peace, that harmony, that can only come with all having the one Ideal; not the one idea, but "Thou shalt love the Lord Thy God with all thine heart, thy neighbor as thyself!" This [is] the whole law, this [is] the whole answer to the world, to each and every soul. That is the answer to the world conditions as they exist today. [Edgar Cayce reading 3976-8]*

The physical universe has a hidden spiritual purpose. A physical foundation of condensed spirit supports an environment of biological organisms joined into a complex, interwoven, organic partnership. Souls incarnating through an advanced life form, a purposely customized human body, are given the opportunity to understand that they have lost awareness of their creator. Through choices that express love in personal relationships, souls can use opportunities presented throughout life to restore their relationship with God. The activity of spirit in the formation of the physical universe has continued for 13.8 billion years, the activity of biological life and the creation of the integrated and diverse biological environment that has spread across the surface of the earth and its oceans has been ongoing for 4.5 billion years, and souls have used human bodies as vehicles for spiritual development for about 106,000 years. How much longer will souls need to incarnate before the last repentant soul finds its way back to God as a perfected independent unit of spirit, mind, and free will? How long can the last recalcitrant souls refuse this amazing opportunity before they forfeit their independent identity and are reabsorbed into the greater Whole because they were unwilling to look beyond selfishness and accept the magnificent future they were offered as companions of the Almighty Creator? Which of these two final destinies will we choose?

THE OPPORTUNITY IS BEFORE THEE! ACCEPT OR REJECT! [Edgar Cayce reading 294-208]

References

Barna Survey, 2011, Six Reasons Young Christians Leave Church: Barna Group, Ventura, California.

Barna Survey, 2020, State of the Church; Signs of Decline & Hope Among Key Metrics of Faith: Barna Group, Ventura, California.

Burke, John, 2015, Imagine Heaven: Baker Books, Grand Rapids, Michigan.

Butler, Anthea, 2020, White Evangelical Racism, The Politics of Morality in America: The University of North Carolina Press, Chapel Hill, North Carolina.

Bob Jones University, 1990, Science 4 for Christian Schools: Bob Jones University Press, Greenville, South Carolina.

Bro, Harmon H., 2011, Edgar Cayce - A Seer Out of Season; The Life of History's Greatest Psychic: A.R.E. Press, Virginia Beach, Virginia.

Cayce E. E., 1968, Edgar Cayce on Atlantis: Paperback Library Inc., New York, New York.

Cayce, H.L., 1980, Earth Changes, Update: A.R.E. Press, Virginia Beach, Virginia.

Chaves, Max, 2019, Inflaton Vacuum Fluctuations as Dark Matter and the Potential V(phi) as Dark Energy: arXiv:1712.07960v4. <https://arxiv.org/abs/1712.07960v4>

Christianity Today, 1992/2021, Why Did So Many Christians Support Slavery?: Christian History magazine, Issue 33, December 2021.

Cole, J.R., Godfrey, L.R., 1985, The Paluxy River Footprint Mystery – Solved: Creation/Evolution Magazine, Vol. 5, No. 1.

Craig, W.L., unknown date, The Teleological Argument - The Fine-Tuning of the Universe: Reasonable Faith.
<https://www.reasonablefaith.org/>

Einstein, Albert, 1905a, On The Movement of Small Particles Suspended in Stationary Liquids Required by the Molecular-Kinetic Theory of Heat: Annals of Physics 17 (1905) 549–560.
<https://einsteinpapers.press.princeton.edu/vol2-trans/137>

Einstein, Albert, 1905b, On a Heuristic Point of View Concerning the Production and Transformation of Light: Annalen der Physik 17 (1905) 132–148.
<https://einsteinpapers.press.princeton.edu/vol2-trans/100>

Einstein, Albert, 1905c, On the Electrodynamics of Moving Bodies: Annalen der Physik 17 (1905) 891–921.
<https://einsteinpapers.press.princeton.edu/vol2-trans/154>

Einstein, Albert, 1905d, Does the Inertia of a Body Depend Upon Its Energy Content?: Annalen der Physik 18 (1905) 639–641.
<https://einsteinpapers.press.princeton.edu/vol2-trans/186>

Einstein, Albert, 1915, On the General Theory of Relativity: Prussian Academy of Science.
<https://einsteinpapers.press.princeton.edu/vol6-trans/110>

Emiliano B., and Osbjorn, P., 2013, Neurocranial Evolution in Modern Humans; The Case of Jebel Irhoud 1: Anthropological Science, Vol. 121, Issue 1, 31–41.

Faccenna, C., et al., 2019, Role of Dynamic Topography in Sustaining the Nile River over 30 Million Years: Nature Geoscience Vol. 12, 1012–1017.

<http://www-udc.ig.utexas.edu/external/becker/preprints/fgfbgsg19.pdf>

Faith, J.T., Surovell, T.A., 2009, Synchronous Extinction of North America's Pleistocene mammals: Proceedings of the National Academy of Sciences Dec 2009, 106 (49) 20641–20645; DOI: 10.1073/pnas.0908153106.

Feynman, R.P., Zee, A., 2014, QED: The Strange Theory of Light and Matter: Princeton Science Library 33, Princeton University Press, Princeton, New Jersey.

Freud, Sigmund, 2010, An Outline of Psycho-Analysis: Martino Fine Books, Eastford, Connecticut.

Fu, Q., Hajdinjak, M., Moldovan, O., et al., 2015, An Early Modern Human From Romania with a Recent Neanderthal Ancestor: Nature 524, 216–219.
<https://doi.org/10.1038/nature14558>

Greene, Brian, 2005, The Fabric of the Cosmos; Space, Time and the Texture Of Reality: Random House Inc., New York, New York.

Greene, Brian, 2010, The Elegant Universe; Superstrings, Hidden Dimensions, and the Quest for the Ultimate Theory: W.W. Norton & Company, Inc., New York, New York.

Greene, Brian, 2011, The Hidden Reality; Parallel Universes and the Deep Laws of the Cosmos: Alfred A. Knopf, New York, New York.

Greyson, Bruce, 2008, Ian Stevenson's Contributions to Near-Death Studies: Journal of Scientific Exploration, Vol. 22, No. 1, 54–63.

Griswold, Eliza, 2021, The Fight for the Heart of the Southern Baptist Convention - How the Convention's Battle Over Race Reveals an Emerging Evangelical Schism: The New Yorker June 10, 2021.

Hamilton, Adam, 2013, When Christians Get It Wrong: Abingdon Press, Nashville, Tennessee.

Hazen, R.M., et al., 2009, Mineral evolution: American Mineralogist,

Vol. 93, 1693–1720. Summary Article - Carnegie Institution, 2008, Mineral Kingdom Has Co-evolved With Life, Scientists Find: Science Daily, 14 Nov 2008.

Herculano-Houzel, Suzana, 2009, The Human Brain in Numbers: A Linearly Scaled-Up Primate Brain: Frontiers in Human Neuroscience 3:31. doi: 10.3389/neuro.09.031.2009.
<https://pmcncbi.nlm.nih.gov/articles/PMC2776484/>

Imes, Jeffrey, 2021, The Essenes and the Advent of Jesus: BookBaby Publishers, 79p.

Jones, R.P., 2020, White Too Long: Simon and Shuster, New York, New York.

Kandel, E. R., 2006, In Search of Memory; The Emergence of a New Science of Mind: W.W Norton and Company, New York, New York.

Larsen, C.S., 2019, Our Origins: W. W. Norton & Company, New York, New York.

Margulis, Lynn and Sagan, Dorion, 1997, Microcosmos, Four Billion years of Microbial Evolution: University of California Press, Berkeley, California.

Martin, Malachi, 1981, The Decline and Fall of the Roman Church: G.P. Putnam's Sons, New York, New York.

Massey R., Kitching T., and Richard J., 2010, The Dark Matter of Gravitational Lensing: Reports on Progress in Physics, arXiv:1001.1739v2.
<https://arxiv.org/abs/1001.1739v2>

Millard, Joseph, 2001, Edgar Cayce: Mystery Man of Miracles: A.R.E. Press, Virginia Beach, Virginia.

Mohr, P. J. and Newell, D. B. and Taylor, B. N., 2016, CODATA recommended values of the fundamental physical constants 2014: Reviews of Modern Physics, Vol. 88, No. 3.
<https://arxiv.org/abs/1507.07956>

Morris, J.D., 1980, Tracking Those Incredible Dinosaurs - And the People Who Knew Them: Master Books.
<https://ncse.ngo/analysis-creationist-film-footprints-stone>

Morrison, L.R., 1981, The Religious Defense of American Slavery Before 1830: Journal of Religious Thought, 37(2):16–29, 1981.

National Centers for Environmental Information, NESDIS, NOAA, U.S. Department of Commerce: Global 3Ma Temperature, Sea Level, and Ice Volume Reconstructions.
<https://www.ncdc.noaa.gov/paleo-search/study/11933>

NASA Exoplanet Archive, Exoplanet and Candidate Exoplanet Statistics.
<https://exoplanetarchive.ipac.caltech.edu/docs/counts_detail.html>

Oesch, P.A., et al, 2016, A Remarkably Luminous Galaxy at z=11.1 Measured with Hubble Space Telescope Grism Spectroscopy: The Astrophysical Journal, 819:129 (11pp), 10 March 2016.

Pew Research Center, 2014, Religious Landscape Study: Pew Research Center, Washington DC.
<https://www.pewresearch.org/religion/religious-landscape-study/views-about-human-evolution/>

Plato, 2008, Timaeus and Critias: Penguin Classics, Penguin Random House, New York, New York.

Principe, L.M., 2006, Science and Religion: The Teaching Company, The Great Courses, No. 4691, Chantilly, Virginia.

Reich, David, 2018, Ancient DNA and the New Science of the Human Past: Evolution Matters Lecture Series, Harvard Museum of Natural History.
<https://youtu.be/990052wQywM>

Reilly, H.J. and Brod, R.H., 1988, The Edgar Cayce Handbook for Health Through Drugless Therapy: A.R.E. Press, Virginia Beach, Virginia.

Schaff, Philip and Wace, Henry, eds., 1900, Second Council of Constantinople: Translated by Henry Percival. From Nicene and Post-Nicene Fathers, Second Series, Vol. 14, Buffalo, NY: Christian Literature Publishing Co., 1900. Revised and edited for New Advent by Kevin Knight.
<http://www.newadvent.org/fathers/3812.htm>.

Stevenson, I. P. and Greyson, Bruce, 1979, Near-Death Experiences, Relevance to the Question of Survival After Death: Journal of the American Medical Association, Vol. 242, 20 Jul 1979.

Stevenson, I., and Keil, J., 2005, Children of Myanmar Who Behave Like Japanese soldiers: A Possible Third Element in Personality: Journal of Scientific Exploration, 19(2), 171–183.

Study Group #1, 2019, A Search for God Book I: Association for Research and Enlightenment, A.R.E. Press, Virginia Beach, Virginia, 61st Printing.

Study Group #1, 2016, A Search for God Book II: Association for Research and Enlightenment, A.R.E. Press, Virginia Beach, Virginia, 35th Printing.

Sugrue, Thomas, 1997, There is a River: A.R.E. Press, Virginia Beach, Virginia.

Summary Article - Hubble Team Breaks Cosmic Distance Record, 3 Mar 2016.
<https://www.nasa.gov/feature/goddard/2016/hubble-team-breaks-cosmic-distance-record>

Susskind, Leonard, 2006, The Cosmic Landscape; String Theory and the Illusion of Intelligent Design: The Hatchette Book Group, New York, New York.

Taylor, S.E., 1973, Footprints in Stone: Eden Films, Films for Christ Association: North Eden Road, Elmwood, Illinois.

The Hidden Mind, April 2002, Scientific American Special Edition,

Vol. 12 No. 1.
<https://www.scribd.com/document/517532895/The-Hidden-Mind>

Todeschi, K.J., 1998, Edgar Cayce on the Akashic Records: A.R.E. Press, Virginia Beach, Virginia.

Todeschi, K.J., 2002, Edgar Cayce on Soul Mates, Unlocking the Dynamics of Soul Attraction: A.R.E. Press, Virginia Beach, Virginia.

Todeschi, K.J., 2014, Edgar Cayce on Vibrations, Spirit in Motion: A.R.E. Press, Virginia Beach, Virginia.

Van Auken, John, 2015, Edgar Cayce's Amazing Interpretation of The Revelation: CreateSpace Independent Publishing Platform.

Vanderburg, Andrew, et al., 2020, A Habitable-Zone Earth-Sized Planet Rescued From False Positive Status: The Astrophysical Journal Letters, Vol. 893, L27, No. 1, 15 Apr 2020. Summary Article by Unknown Editors, 2005, Earth Size, Habitable-Zone Planet Found Hidden in Early NASA Kepler Data: Jet Propulsion Laboratory, California Institute of Technology 15 Apr 2020.
<https://www.jpl.nasa.gov/news/earth-size-habitable-zone-planet-found-hidden-in-early-nasa-kepler-data>

Villaver, Eva and Livio, Mario, 2007, Can Planets Survive Stellar Evolution?: The Astrophysical Journal, 661:1192–1201, 1 June 2007.

Welker, Frido, et al., 2020, The Dental Proteome of Homo Antecessor: Nature; DOI: 10.1038/s41586-020-2153-8.

Wellcome, 2018, Wellcome Global Monitor Survey: Wellcome, London, United Kingdom.
<https://wellcome.org/reports/wellcome-global-monitor/2018>

Whittle, M., 2008, Cosmology; The History and Nature of Our Universe: The Teaching Company, The Great Courses, no. 1830, Chantilly, Virginia.

Willner, John, 2001, The Perfect Horoscope: Following the Astrological Guidelines Established by Edgar Cayce: Cosimo Inc., Paraview

Press, New York, New York.

Xue, Chao, et al., 2020, Precision Measurement of the Newtonian Gravitational Constant: National Science Review, Volume 7, Issue 12, December 2020, 1803–1817.
<https://doi.org/10.1093/nsr/nwaa165>

Scientific Notation

Scientists often deal with very large and very small numbers that can be extremely cumbersome to write in decimal form. Scientific notation is used to allow these numbers to be written in a more compact form as the product of a smaller decimal number and a power of ten; for example, 1.1056×10^{9} (voiced as one point one zero five six times ten to the ninth) is the number 1,105,600,000. The powers of ten can be negative as well as positive; for example, 1.1056×10^{-9} (voiced as one point one zero five six times ten to the minus ninth) is the number 0.0000000011056. If the decimal part is 1, it is usually not written; thus, 1×10^{9} is written as 10^{9} (voiced as ten to the ninth) and is the number 1,000,000,000 (one billion), or 1 followed by nine zeros. Similarly, 1×10^{-9} is written in abbreviated form as 10^{-9} (voiced as ten to the minus ninth) and is the number 0.000000001 (one billionth), or a decimal point followed by 8 zeros and a 1. Scientific notation can greatly simplify arithmetic calculations and allows easy comparison of numbers.

Edgar Cayce And The A.R.E.

The Edgar Cayce readings are housed at the Association for Research and Enlightenment (A.R.E.) located in Virginia Beach, Virginia. The A.R.E. not only archives the readings and associated correspondence to and from the people who requested readings but also publishes, teaches, and explores the concepts presented in the readings. The topics of investigation include holistic health, meditation and prayer, spirituality and the relationship of man and God, dream interpretation and intuition, psychic phenomena, and reincarnation. Many of the readings offer insights into humankind's ancient history and long struggle to find its way back to God. The A.R.E. actively promotes research into various topics of the readings and organizes conferences and webinars to disseminate information from the readings and related topics. The affiliated Atlantic University offers courses in transpersonal psychology and mindfulness using concepts from the readings to study the inter-relationships among body, mind, and spirit. The university stresses the importance of service to humanity and the need for each individual to contribute to the spiritual advancement of humankind.

References to the Edgar Cayce readings are denoted by a sequence of (usually) two digits. For example, the number 5749-8 refers to reading number 5749 (a number assigned to the person for whom the reading was conducted to ensure anonymity) and the eighth reading in a series of readings for that person. A single digit reference enclosed in brackets refers to the person assigned that number.

About the Author

Jeffrey Imes is a retired PhD physicist. His career included working with a private geophysics firm to locate oil reservoirs in Libya and Tunisia, for the U.S. Geological Survey (USGS) to search for offshore oil reservoirs in the Gulf of Mexico, and with the USGS to develop computer simulations of groundwater flow through the aquifers of southern Missouri. He also studied the aquifers of Abu Dhabi Emirate, worked with the U.S. Navy and USAID to identify promising locations for village water wells in Kenya and Djibouti, and worked with the World Health Organization to study aquifer production and arsenic contamination in Bangladesh. His interests range from cosmology to archaeology to spirituality. He has studied the Edgar Cayce readings for about 45 years and appreciates the natural way in which the philosophy of the readings integrate science, Christianity, and certain aspects of Eastern religious thought in a coherent non-dogmatic manner. He has written two other books that explore Christianity and the nature of the soul from the perspective of the Edgar Cayce Readings: Consciousness, Reality, and the Spiritual Journey of the Wayward Soul (2026) and The Essenes and the Advent of Jesus (2021).

Learn more at jeffreyimes-author.com.

www.ingramcontent.com/pod-product-compliance
Lightning Source LLC
Chambersburg PA
CBHW060515160726
47991CB00001B/50